# DOWNSTREAM CONSEQUENCES OF DAMMING, LAKE TANA BASIN, ETHIOPIA

Chalachew Abebe Mulatu

# DOWNSTREAM CONSEQUENCES OF RIBB RIVER DAMMING, LAKE TANA BASIN, ETHIOPIA

## DISSERTATION

Submitted in fulfillment of the requirements of

the Board for Doctorates of Delft University of Technology

and

of the Academic Board of the IHE Delft

Institute for Water Education

for

the Degree of DOCTOR

to be defended in public on

Thursday, 16 December 2021, 15:00 hours

in Delft, the Netherlands

by

Chalachew Abebe MULATU

Master of Science in Water Science and Engineering; Specialization Hydraulic

Engineering and River Basin Development

IHE Delft Institute for Water Education, Delft, the Netherlands

born in Dangila, Ethiopia

This dissertation has been approved by the

promotor: Prof.dr. M.E. McClain and
copromotor: Dr.ir. A. Crosato

Composition of the doctoral committee:

| | |
|---|---|
| Rector Magnificus TU Delft | Chairman |
| Rector IHE Delft | Vice-Chairman |
| | |
| Prof.dr. M.E. McClain | IHE Delft / TU Delft, promotor |
| Dr. ir. A. Crosato | IHE Delft, copromotor |
| | |
| Independent members: | |
| Prof.dr. N.C. van de Giesen | TU Delft |
| Dr.ir. C.M.M. Mannaerts | University of Twente |
| Prof. T. Steenhuis | Cornell University, USA |
| Dr.L. Brandimarte | KTH Royal Institute of Technology Stockholm, Sweden |
| | |
| Prof.dr. C. Zevenbergen | TU Delft / IHE Delft, reserve member |

*This research was conducted under the auspices of the Graduate School for Socio-Economic and Natural Sciences of the Environment (SENSE)*

CRC Press/Balkema is an imprint of the Taylor & Francis Group, an informa business

Published by:
CRC Press/Balkema
enquiries@taylorandfrancis.com
www.crcpress.com – www.taylorandfrancis.com
ISBN 978-1-032-25031-1

# ACKNOWLEDGMENTS

This thesis work would not have come to end without the support and encouragement of different individuals and organizations. First, I gratefully like to acknowledge my esteemed promoter, Prof. dr. Michael E. McClain, for his willingness and allowing me to do my study under his supervision. Without his exceptional support and encouragement, I would not finalize this research work. It may be hard to forget his motivation, trust, support, and respect in writing the articles and the dissertation. He is a great scientific father that understand me very well and respond fruitfully in a friendly manner. Prof., believe me, the word "thank you" is not enough to you.

I also would like to express my sincere thanks to my co-promotor, Dr. Alessandra Crosato, for her willingness to work with me, critical comments, suggestion and valuable discussion throughout the research period. I appreciate her for teaching me how to do a research and disseminate the findings in short and clear manner. Without her guidance and support, it would be impossible to finish the work.

This work would also be impossible without the help, consultation, and discussion of Dr. Eddy J. Langendoen, Research Hydraulic Engineer in Agricultural Research Service, U.S. Department of Agriculture. I would like acknowledge his unreserved help throughout this research work. I would also like to thank Dr. Michael M. Moges, Associate professor in the Faculty of Civil and Water Resources Engineering, Bahir Dar University, Ethiopia, for his help as a colleague and as a Ph.D. research committee. I would like to thank Dr. Kees Sloff (Assistant Professor at Delft University of Technology and Expert on fluvial river engineering and fluvial morphology at Deltares, Delft, The Netherlands) for his valuable support to setup SOBEK-RE model and, Dr. Seleshi Yalew (Postdoc Researcher at TU Delft) for helping me to write the Google Earth engine script.

I would also like to express my sincere thanks to Prof. dr. ir. Arthur Mynett for his willingness to be my promoter at the start of the work. Officially, I was registered as his Ph.D. student before he retires. I thank him for his careful guidance and support in developing my Ph.D. research proposal. I also thank all the doctoral committee for their time to read the thesis work, and valuable comments and suggestions.

I thank the financial support of the Netherlands Fellowships Programme (NFP) to undertake this study. I also thank J. Boots and A. Karsten for their help in managing the financial and administrative issues. I am thankful to Bahir Dar University Technology of Institute for allowing me to travel to the Netherlands to do my study. I acknowledge the discussion and support of IHE Ph.D. fellows, Ethiopian Ph.D., and MSc students. I would like to thank Dr. Kedir and Meles for making my life easy in the Netherlands. My thanks also go to Dr. Fenta, Bitew, Wubneh, Goraw, Elias, and many others who are not mentioned here.

I thank my beloved wife, Abebech Ademe, for her love, support, understand and care of my children in my absence. Abe, I thank you for your patience and encouragement and, strength to carry all responsibilities. I also thank my lovely kids: Absira, Mahider, and Hana for their love and inspiration. You all always give me positive energy on my life. My heart-full appreciation goes to my father, mother, brothers, sisters, and family members for their encouragement, moral support, and prayer. I thank Joy Pengel for helping me to translate the summary to Dutch. Last but not least, I would like to thank the almighty GOD and his beloved mother for helping me to accomplish this long journey.

Chalachew Abebe Mulatu,

Delft, December 2021

# SUMMARY

The primary effects of dams on downstream rivers are hydrological alterations, generally characterized by reductions of discharge peaks and increases of low flows, together with changes in their duration, occurrence period, and frequency. By blocking the movement of sediment produced in the upstream watershed, dams drastically reduce the sediment input to the downstream river. As a consequence, the river gradually adjusts its shape and slope based on released discharges, amounts, and type of sediment bypassing the dam, as well as geological setting. The reduction of peak discharges has a mitigating effect on downstream flooding, but this may also have important consequences on the ecological functioning of river channel and floodplain ecosystems.

The Lake Tana Basin is part of the Upper Blue Nile Basin in the north-western of Ethiopia. It is one of the country's major development corridors for its water resource potential and suitable sites for hydropower and irrigation. Planned dams in the basin are mainly for irrigation projects on the flat lands around Lake Tana and. to prevent flooding on downstream plains, which is one of the major problems of the basin during the rainy season. The Ribb Dam irrigation is one of the projects currently under construction, with a dam and a diversion weir at 77 km and 50 km upstream of the Ribb River mouth, respectively. The project will irrigate 15,000 ha of the Fogera Plain, where the Ribb enters Lake Tana. The Fogera Plain is also influenced by flooding from the Gumara River, which flows parallel to the Ribb River to the South.

This study analyze how the Ribb River system may adapt to the Ribb Dam construction. The study includes analyses of (i) the morphological trends of the river system before damming, (ii) the pre- and post-dam discharge and sediment transport regimes, (iii) the long-term morphological effects of different dam operation scenarios, and (iv) the effects of the dam on the Fogera Plain, considering also its ecological functioning. The study combines primary and secondary data collection, the application of remote sensing, and mathematical modeling.

The morphodynamic trends of the Ribb River before damming are analyzed for 59 years (1957 to 2016) based on aerial photographs, SPOT and Google Earth satellite images, and collected field data. The analysis shows that the river system experienced channel cut-off and avulsion, resulting in reduced channel length. The avulsion was caused by the blockage of the old river channel due to excessive sedimentation, creating a 16.4 km new waterway draining to Lake Tana. The reach-averaged sediment transport capacity of the river system is estimated using empirical sediment transport equations after dividing the downstream river channel into four reaches based on natural and manmade channel bed features, anthropogenic interventions, and channel slope and width. Results indicate reduced sediment transport capacity downstream. The river channel width also reduces downstream due to continuous sediment deposition.

The downstream river system is influenced by anthropogenic activities like sand mining, water withdrawal for irrigation, embankment construction for flood protection, and regulation of Lake Tana level for hydropower production. The effects of these activities are analysed as part of this study. Continuous sand mining is found to result in reduced longitudinal river bed slope, as analysed using the morphodynamic Equilibrium Theory. Instead, dry season water extraction for irrigation is found to have negligible effects on the river bed slope, as was the effect of Lake Tana water level regulation.

The long-term hydro-morphological effects of the Ribb Dam on the downstream river reaches are assessed for different dam operation scenarios to select the least-impacting one and to assess the applicability of a simplified analytical method for the quick assessment of dam induced morphological impacts. The Ribb Dam operation will increase and decrease the dry- and wet-season discharges, respectively, and will shift the timing of low and high discharges. The change in discharge regime will reduce the sediment transport capacity of the downstream river reaches compared with the pre-dam situation. In response, the river will adjust its channel geometry by reducing its slope as predicted by the morphodynamic Equilibrium Theory and 1D mathematical model.

The Fogera Plain experiences recurrent flooding due to the overbank flow of the Ribb and Gumara Rivers. Flooding affects infrastructures like roads, houses, health centres, and threatens human lives. The HEC-HMS model was used to analyze the capability of the Ribb Dam to reduce flooding on the Fogera Plain. The simulations show that the dam outflow peak discharge will be reduced on average by 20%. However, the post-dam discharge value at the Lower Gauging station near Fogera Plain does not show significant changes due to discharge contributions from the downstream watershed between the dam and the station. The model is also used to generate the discharge time-series of the rivers as the existing gauging stations are located in the floodplain, where the peak discharge that causes flooding is not measured due to reduced channel dimensions and overbank flow. The generated time-series is then used to characterize the pre- and post-dam regimes of the Ribb River and to study the flooding extent over the plain.

The effect of Ribb Dam on the flooding extent of the Fogera Plain is assessed using the HEC-RAS 2D hydrodynamic model. The results show that the dam may reduce the flooding extent by 11% on average. As the historical flooding extent of the area was unavailable to calibrate the hydrodynamic model, calibration is based on the spectral reflectance difference of Landsat satellite images using the Google Earth Engine. This method allowed analyzing multiband satellite imagery to obtain a grid-based inundation map for the selected time frame. Model calibration show good agreement with the retrieved satellite image of the Google Earth Engine, with a measure of fit of 52%. The finding show that the Ribb Dam operation will have a small effect on flooding extent, depth, and flood duration of the Fogera Plain. This indicates that hydrological alterations to the plain should be small also regarding the ecological processes. However, the construction of irrigation structures on the floodplain may affect the free movement of water to the wetlands and fish migration. Moreover, the agricultural activities will reduce available habitats for the aquatic community and, endemic and migratory birds. The probable increase of the dry-time post-dam release of drainage water from the farmlands to the wetlands may prolong the water availability period of the wetlands and provide some support to the aquatic community if the water quality is maintained.

The analysis of dam-induced hydrological and morphological impacts on downstream river systems is important for better understanding the impacts, and especially for the design of mitigation strategies. The results of this research contribute to the expanding global knowledge base regarding the environmental impacts of dams on low-land river systems, as well as specific and highly relevant knowledge about impacts in the Lake Tana Basin of Ethiopia. The developed methodologies and findings may be used as the base-case for the study of future hydro-morphological and ecological changes in the river system that may arise due to other dam operations or climate change.

# SAMENVATTING

De primaire effecten van dammen op benedenstroomse rivieren zijn hydrologische veranderingen, die doorgaans worden gekenmerkt door; afname van afvoerpieken en toename van lagere waterstanden, samen met veranderingen in hun duur, hoe vaak ze voorkomen en hun frequentie. Door het blokkeren van sediment transportatie van de stroomopwaartse gebieden, verminderen dammen de toevoer van sediment naar de stroomafwaartse rivier drastisch. Als gevolg hiervan past de rivier geleidelijk zijn vorm en helling aan op basis van hoeveelheid waterafvoer, de hoeveelheden en type sediment dat de dam omzeilt, evenals de geologische omgeving. Het terugdringen van piekafvoeren verminderd de kans op benedenstroomse overstromingen. Maar het tegengaan van overstromingen kan ook ernstige gevolgen hebben voor het ecosysteem van de riviergeulen en uiterwaarden.

Het Lake Tana Basin maakt deel uit van het Upper Blue Nile Basin in het noordwesten van Ethiopië. Het is een van de belangrijkste ontwikkelingscorridors van het land vanwege de potenties die er mogelijk zijn zoals; waterberging, Hydro power en irrigatie. Geplande dammen in het stroomgebied zijn voornamelijk bedoeld voor irrigatieprojecten op het vlakke land rond het Tanameer. Ook helpt het overstromingen stroomafwaarts te voorkomen, wat een van de grootste problemen is in die gebieden tijdens het regenseizoen. De Ribb Dam-irrigatie is een van de projecten die momenteel in aanbouw is. Hier komt een dam en een omleidingsstuw op respectievelijk 77 km en 30 km stroomopwaarts van de Ribb-riviermonding. Het project zal irrigatie mogelijkheden creëren voor 15.000 ha van de Fogera-vlakte, gelegen in de uiterwaarden van de rivier de Ribb nabij zijn monding. De Fogera-vlakte wordt ook beïnvloed door overstromingen van de Gumara-rivier, die parallel aan de Ribb-rivier naar het zuiden stroomt.

Deze studie analyseert hoe het Ribb-riviersysteem zich mogelijk aanpast aan de Ribb-damconstructie. De studie omvat analyse van (i) de morfologische trends van het riviersysteem vóór de afdamming, (ii) de regimes voor sedimenttransport en plaatsing voor en na de dam, (iii) de morfologische effecten op lange termijn van verschillende scenario's voor de werking van dammen, en (iv) de effecten van de dam op de Fogera-vlakte, gezamenlijk met de ecologische werking ervan. De studie combineerde primaire en secundaire gegevensverzameling, de toepassing van teledetectie en wiskundige modellering.

De morfodynamische trends van de Ribb-rivier vóór de afdamming worden werden gedurende 59 jaar (1957 tot 2016) geanalyseerd op basis van luchtfoto's, SPOT- en Google Earth-satellietbeelden en recentelijk verzameld veldgegevens. Uit de analyse bleek dat het riviersysteem kanaalafsnijding en -avulsie vertoonde, wat de kanaallengte verkleind. De avulsie werd veroorzaakt door de blokkering van het oude rivierkanaal als gevolg van sedimentatie, wat een nieuwe waterweg van 16,4 km creëerde naar het Tanameer. De bereik-gemiddelde sedimenttransportcapaciteit van het riviersysteem werd geschat met behulp van empirische sedimenttransport berekeningen. Hiervoor werd stroomafwaarts het kanaal in vier bereiken was verdeeld op basis van; natuurlijke en door de mens veroorzaakte kanaalbodemkenmerken, antropogene ingrepen, en kanaalhelling en -breedte. De resultaten wezen op een verminderde transportcapaciteit van sediment stroomafwaarts. De rivierkanaalbreedte neemt ook stroomafwaarts af als gevolg van continue sedimentafzetting.

Het stroomafwaartse riviersysteem wordt beïnvloed door antropogene activiteiten zoals zandwinning, wateronttrekking voor irrigatie, aanleg van dijken voor waterkering en regulering van het Tanameer-niveau voor Hydropower. In het kader van dit onderzoek zijn de effecten van deze activiteiten geanalyseerd. Continue zandwinning in het Middle Reach bleek te resulteren in een verminderde longitudinale rivierbedding. Dit is geanalyseerd met behulp van de morfodynamische evenwichtstheorie. Wateronttrekking in het droogseizoen voor irrigatie bleek een verwaarloosbaar effect te hebben op de veranderingen in de helling van de rivierbedding, evenals het reguleren van het waterpeil in het Tanameer.

De hydromorfologische langetermijneffecten van de Ribb-dam op het stroomafwaartse rivierbereik werden beoordeeld voor verschillende scenario's voor de werking van dammen om het scenario met de minste impact van de dam te selecteren. Maar ook om de toepasbaarheid te beoordelen van een vereenvoudigde analysemethode voor een snelle beoordeling van door dam veroorzaakte morfologische effecten in kaart te brengen. De Ribb Dam-operatie zal het verschil in de waterafvoer in het droge en natte seizoen veranderen. Ook zal de timing van lage en hoge afvoer verschuiven. Door de wijziging van het afvoerregime zal de sedimenttransportcapaciteit van het stroomafwaarts rivierbereik afnemen in vergelijking met de situatie vóór de dam. Hierdoor zal het rivierbereik zijn kanaalgeometrie aanpassen door zijn helling te verminderen, zoals de morfodynamische evenwichtstheorie en het 1D-wiskundig model voorspelt.

De Fogera-vlakte ervaart terugkerende overstromingen als gevolg van de oeverstroming van de Ribb- en Gumara-rivieren. Overstromingen hebben gevolgen voor infrastructuren zoals wegen, huizen, gezondheidscentra en bedreigen voor mensenlevens. Het HEC-HMS-model werd gebruikt om het vermogen van de Ribb-dam om overstromingen op de Fogera-vlakte te verminderen te analyseren. Uit de simulatie bleek dat de piekafvoer van de damuitstroom gemiddeld met 20% zal worden verminderd. De waarde van de afvoer na de dam bij het Lower Gauging-station nabij de Fogera-vlakte vertoont echter geen significante veranderingen. Dit komt voornamelijk vanwege de afvoerbijdragen stroomafwaarts tussen de dam en het station. Het model is ook gebruikt om tijdreeksafvoerwaarden van de rivieren te genereren, aangezien de bestaande meetstations zich in de uiterwaarden bevinden waar de piekafvoer die overstromingen veroorzaakt niet wordt gemeten vanwege het verkleinen van het kanaal. De gegenereerde tijdreeksafvoerwaarde wordt gebruikt om de gecombineerde pre- en post-damafvoercondities van de Ribb-rivier en de mate van overstroming van de vlakte te karakteriseren.

Het effect van Ribb Dam op het overstromingsgebied van de Fogera Plain werd beoordeeld met behulp van het HEC-RAS 2D hydrodynamische model. Hieruit bleek dat de dam de overstromingsgraad met gemiddeld 11% kan verminderen. Omdat de historische overstromingsomvang van het gebied niet beschikbaar was om het hydrodynamische model te kalibreren, werd de kalibratie uitgevoerd op basis van het spectrale reflectieverschil van Landsat-satellietbeelden met behulp van de Google Earth Engine. Met deze methode konden multiband-satellietbeelden worden geanalyseerd om een op rasters gebaseerde overstromingskaart voor het geselecteerde tijdsbestek te verkrijgen. Modelkalibratie toonde een goede overeenkomst met het opgehaalde satellietbeeld van de Google Earth Engine van 52%. De bevinding dat de Ribb Dam-operatie een klein effect zal hebben op de mate van overstroming, diepte en overstromingsduur van de Fogera-vlakte geeft aan dat hydrologische veranderingen in de vlakte ook minimaal moeten zijn voor ecologische processen. De aanleg van irrigatiestructuren op de uiterwaarden kan echter het vrije verkeer van water naar wetlands en vismigratie nadelig beïnvloeden, en landbouwactiviteiten verminderen de beschikbare habitats voor de aquatische gemeenschap en inheemse en trek-vogels. De waarschijnlijke toename van de droogtijd, na het vrijkomen van de dam van drainagewater van de landbouwgronden naar de wetlands, kan de waterbeschikbaarheidsperiode van de wetlands verlengen en kan de watergemeenschap enige steun bieden zolang de waterkwaliteit behouden blijft.

De analyse van door dammen veroorzaakte hydrologische en morfologische effecten op stroomafwaartse riviersystemen is belangrijk voor een beter begrip van de effecten en het ontwerpen van mitigatiestrategieën. De resultaten van dit onderzoek dragen bij aan de groeiende wereldwijde kennis van de sociale en milieueffecten van dammen op laaglandriviersystemen, evenals aan specifieke en zeer relevante kennis over de effecten in het Tanameer-bekken in Ethiopië. De ontwikkelde methodologieën en bevindingen kunnen worden gebruikt als basisscenario voor toekomstige studies over mogelijke hydromorfologische en ecologische veranderingen in het riviersysteem die kunnen optreden als gevolg van andere damoperaties of vanege klimaatverandering.

# CONTENTS

# 1

# INTRODUCTION

This chapter briefly explains the background, the aim and the approach of this PhD study. The general structure of the document is also included.

## 1.1 BACKGROUND

The increased water demand for irrigation, power generation, domestic and industrial supplies in Ethiopia requires proper planning and management of the resources. In 2002, the Ethiopian Ministry of Water, Irrigation, and Energy (MoWIE) produced a water sector development program for the short (2002-2006), medium (2006-2011), and long-term (2011-2016) to construct large, medium and small-scale projects for irrigation and hydropower. The program planned to irrigate an additional area of 1.33, 1.56 and 1.81 million ha in the short, medium and long-terms, respectively (MoWR, 2002). Power production was also planned to increase from 1,314 GWh/year in 2002 to 2,003, 2,840 and 4,040 GWh/year in the short, medium and long-term plans, respectively (MoWR, 2002). One example of the projects to achieve this strategy is the construction of the Grand Ethiopian Renaissance Dam (GERD), which was started in 2011 to store 74 Bm$^3$ of water for hydropower. The project is planned to generate 6,500 Mega Watts, which makes it the largest hydropower plant in Africa. Moreover, a number of water resource development projects are either under construction or planned in different parts of the country.

The government of Ethiopia identifies the Lake Tana Basin (Figure 2.1C), part of the Upper Blue Nile Basin, as one of the developmental areas because of its water resources and fertile land potential. Consequently, a number of projects are under construction for irrigation and hydropower. This includes the Ribb Dam on the Ribb River and the Megceh Dam on the Megech River to store 234 Mm$^3$ and 181 Mm$^3$ of water to irrigate 15,000 ha and 7,300 ha, respectively (BRLi and MCE, 2010).

Dams affect the hydro-morphological process of the downstream river reaches by storing water and sediment (Graf, 2006; Ronco et al., 2010; Marcinkowski and Grygoruk, 2017; Sanyal, 2017). The effect is highly dependent on the size of the reservoir and the purpose of operation and, considerably varies between them whereby one may release very little and the other may barely alter the flow (Magilligan and Nislow, 2005; Childs, 2010). However, all are expected to have some impacts on the downstream (Baxter, 1977; Williams and Wolman, 1984; Brandt, 2000; Childs, 2010). The ratio of sediment storage to the total inflow by the reservoir (trap efficiency) mainly depends on parameters such as storage capacity, incoming sediment volume and size, sediment outflow, outlets and type of control structures (Brune, 1953). The study done by Williams and Wolman (1984) indicated that large dams may store almost all the incoming sediment volume with a trapping efficiency greater than 99.5%. This shows that for large dams, almost all the coarser sediments are deposited in the reservoir and only a small fraction of suspended sediment passes through the outlets (Brandt, 2000). This, in general, results in reduced sediment concentration and smaller sediment sizes in the downstream river reaches (Kondolf, 1997).

Dams also reduce the downstream flooding and affect the river, wetlands, and riparian ecosystems, ground water conditions, vegetation cover, etc (Batalla et al., 2004; Petts and Gurnell, 2005). Flood reduction affects fish migration, plant habitat, water quality, and nutrient inputs (Lytle and Poff, 2004), as well as ecologically important flow variables (ecological thresholds) (Marcinkowski and Grygoruk, 2017; Teng et al., 2017). These include the alteration of the natural river discharge dynamics, sediments, and biodiversity (Syvitski et al., 2005; Biemans et al., 2011) which may provoke invasion by exotic species (Poff et al., 1997).

The reduction of water discharge and sediment load produces geomorphic effects on the downstream river channel (Kondolf, 1997; Phillips et al., 2005). The river may respond to the new intervention with bed aggradation or degradation, channel narrowing or widening, depending on the sediment transport capacity of the flow (Williams and Wolman, 1984; Osterkamp et al., 1998; Petts and Gurnell, 2005; Graf, 2006). The size of the river and the dam, the pre- and post-dam hydrologic regimes, the characteristics of riparian vegetation, the environmental setting, the initial channel morphology, the sediment inputs and the type of released sediments, and the operation schedule may affect the extent of downstream morphological impacts and the required time (relaxation period) for the geomorphological adjustments to reach a new equilibrium (Williams and Wolman, 1984; Lagasse et al., 2004; Petts and Gurnell, 2005; Yang et al., 2011). In the morphological adaptation period, the discharge released from the dam may erode the river bed and banks until equilibrium between sediment transport capacity and sediment input is attained, which generally results in a reduced river bed slope or river channel widening (Williams and Wolman, 1984; Childs, 2010).

The dam-induced geomorphological effects on the downstream reaches have been studied by a number of geomorphologists (Williams and Wolman, 1984; Huang and Nanson, 2000; Phillips et al., 2005; Graf, 2006; Curtis et al., 2010; Kondolf et al., 2014a). They have developed empirical, conceptual, and predictive methods for both case-related and generic assessments (Grant, 2012). In addition, one, two and three-dimensional mathematical models have been developed to simulate and analyze the short, intermediate and long-term dam-induced impact (Khan et al., 2014; Omer et al., 2015). A general method to determine dam-induced impacts on downstream channel morphology is lacking due to the existence of complex interactions between the site-specific drivers and processes (Williams and Wolman, 1984; Magilligan and Nislow, 2005; Childs, 2010).

In the Lake Tana Basin of Ethiopia, different water resources projects for irrigation, hydropower, and water supply are either constructed, under construction, or planned. These developmental projects will strongly affect the water resource, sediment transport and, ecological conditions of the rivers. However, insufficient attention is given in the basin to analyze the adaptation mechanism of river systems downstream of dams. Hence, this study deals with the hydrological, morphological, and ecological consequences of the Ribb Dam construction.

## 1.2 RESEARCH AIMS

The main objective of this study is to obtain an improved understanding of the adaptation of the Ribb River system, located in Ethiopia, in response to upstream damming (Figure 2.1D). The research has the following specific objectives:

1. To analyze the morphological trends of the Ribb River before dam construction,

2. To analyze the long-term morphological effects of different dam operation scenarios,

3. To investigate the applicability of the simplified analytical method for quick assessment of morphological impacts on downstream river reaches, and

4. To analyze the capability of the Ribb Dam to change the Fogera Plain flooding including some implications for the ecology of the floodplain wetlands.

## 1.3 RESEARCH QUESTIONS

This work answers the following major research questions to achieve the developed objectives. It includes:

1. What were the morphological trends of the Ribb River before damming?

2. How will the Ribb Dam operation affect the river discharge regime and the Fogera Plain flooding extent?

3. What are the major factors related to flooding that influence the Fogera Plain ecology and to what extent will the operation of the Ribb Dam affect them? and

4. How will dam operation affect the morphological condition of the downstream river system on the long term?

## 1.4 GENERAL APPROACH

The approach of this research includes detailed literature reviews on the effects of dam construction on the hydro-morphology of rivers and on the application of state-of-the-art numerical and mathematical modeling to assess dam-induced impacts.

The Ribb River is selected as the case study site, as a large dam is under-construction in its upper reach to store water for irrigation. The research includes the acquisition of primary field data including bed material grain size, reach-averaged river bed slope and channel width from the identified study river reaches. Secondary data, such as discharge time-series, water levels of Lake Tana and along the Ribb River at the gauging stations are collected from Ministry of Water, Irrigation and Energy (MoWIE). Rainfall and other climatic data at the representative meteorological stations of the study watersheds are derived from the Ethiopian National Meteorological Agency (NMA).

Based on literature and data, the study identifies the nature and the rate of river channel changes (morphological trends), and their natural and anthropogenic drivers before dam construction. This is needed to identify and understand the effects of the Ribb Dam on the Ribb River. This part of the work is based on data analysis and on the application of the Equilibrium Theory of Jansen et al. (1979). The study includes the analysis of retrogressive propagation of bed adaptation due to changes in downstream boundary conditions (Lake Tana levels) and the contribution of these changes to siltation in the lower river reach. The timing of bed-level change propagation is derived based on the theory developed by De Vries (1975).

The study also determines the pre- and post-dam hydrological regimes of the downstream river reaches which are necessary to compute the dam-induced morphological changes. The Equilibrium Theory of Jansen et al. (1979) is applied to the sand- and gravel-bed river reaches to assess whether the theory can be used for quick assessments of long-term morphological effects of different dam operations. For this, the results of the theory, describing the final morphological state of the river, are compared with the results of a calibrated 1D morphological model for the same boundary conditions. The time-scale of the river reach to achieve the theoretical final state of morphological equilibrium is established based on the results of the 1D model.

Finally, the study assesses the effects of the dam on extent and duration of flooding over the Fogera Plain and the wetlands (Figure 2.1D) which are home for endangered and endemic birds and nursery sites for different fish species. A HEC-HMS hydrological model is developed for the main rivers (the Ribb and the Gumara) that pass through the Fogera Plain to analyze the pre- and post-dam discharge time series at the lower gauging stations, as the peak values that cause flooding are not measured there due to reduced channel conveyance capacity. These discharges are then used as inputs for the 2D hydrodynamic model of HEC-RAS to assess the pre- and post-dam flooding. Historical inundation maps are retrieved using Landsat satellite image reflectance values and used to calibrate the 2D hydrodynamic model based on flooding extent. The difference between pre- and post-dam flooding extent, duration, depth, due to the alteration of hydrological conditions by the dam and the effects of irrigation infrastructures on the floodplain, are considered to assess the ecological consequences for the Fogera Plain wetlands.

## 1.5 STRUCTURE OF THE THESIS

The thesis includes six chapters. A short overview of each chapter is presented below.

**Chapter 1** includes the introduction describing briefly the study background, the research aims, and the general approach of the work.

**Chapter 2** describes the Lake Tana Basin, the study watersheds and the river reaches. The geographical location, the hydrological and climatological conditions, the past, present, and future anthropogenic activities (interventions), the land use and land cover (LUCL) conditions, the water resource potentials, and major water resource developments are discussed.

**Chapter 3** details a quantitative analysis of the morphodynamic trends of the Ribb River before damming. The effects of anthropogenic activities like water extraction for irrigation, sand mining, embankment construction, and Lake Tana water level regulation on the river morphological changes are examined. The results are interpreted in terms of current understanding of sand-bed rivers. The analysis helps to draw a baseline to study the morphological changes of the downstream river reaches due to the Ribb Dam construction.

**Chapter 4** presents the long-term hydro-morphological changes of the Ribb River system downstream of the dam. The work is also meant to investigate the applicability of the simplified analytical method for quick assessments of the long-term morphological effects on the downstream river reaches. A 1D SOBEK-RE model is used to compare the results and estimate the required morphological time of the river system to reach the theoretical final state of equilibrium.

**Chapter 5** presents the effect of the Ribb Dam operation on the discharge regime of the river and on the Fogera Plain flooding extent highlighting the consequences for the floodplain wetlands ecology.

**Chapter 6** presents the general discussion and conclusions providing the answers to the research questions addressed by this study, the scientific contribution of the study, and the limitations for future improvements.

# 2
# THE LAKE TANA BASIN

This chapter presents the topography, the land use and land cover, the rainfall and climatic conditions and the developed, ongoing and planned water resources projects of the Lake Tana Basin. Moreover, a detailed description of the Ribb and Gumara watersheds and the Fogera Plain is presented.

## 2.1 DESCRIPTION OF THE BASIN

The Lake Tana Basin (Figure 2.1C) is part of the Upper Blue Nile Basin (UBNB) (Figure 2.1B) in the north-western part of Ethiopia, at a latitude of 10°58′–12°47′N and a longitude of 36°45′–38°14′E, with a surface area of 15,000 km². The UBNB covers 17.5% of Ethiopia's surface area with an estimated average annual runoff of 54.8 billion m³ contributing for 50% and 40% to the country's total average annual runoff and agricultural production, respectively (Awulachew et al., 2007). Lake Tana is the biggest freshwater lake in Ethiopia. With an estimated surface area of 3,000 km² and maximum and average depths of 14 and 9 m, respectively, the water residence time of the lake is 1.5 years (SMEC, 2008b). Lake Tana is fed by more than 40 rivers and streams (McCartney et al., 2010) and it is the source of the Blue Nile River, the only natural outflow that contributes 8% to the Nile River flow (Conway, 1997; Kebede et al., 2006). More than 93% of the discharge to the Lake Tana is produced by the four major rivers: namely, the Ribb, the Gumara, the Megech, and the Gilgel Abay (Kebede et al., 2006). The total inflow from the above rivers is estimated in 6,788 Mm³/year, while the outflow to the Blue Nile River and the average annual evaporation from the Lake surface are 4,789 Mm³/year and 5,138 Mm³/year, respectively (Duan et al., 2018).

Lake Tana is an important economic resource of the area including transportation, lakeside agricultural activities (with and without irrigation), sand mining, domestic and industrial water supply, and hydroelectric power production. Fishing in the area gives socio-economic benefits in terms of employment, income generation, nutritional values, and food security (Mengistu et al., 2017). The lake and its surroundings are a touristic attraction due to the existence of ancient Ethiopian Orthodox Churches and Monasteries in the 37 islands and 16 peninsulas (McCartney et al., 2010) and a 40 m high waterfall (Tis Issat Fall) on the Blue Nile River 35 km downstream from Bahir Dar (Vijverberg et al., 2009). Moreover, the basin has been recognized as a Biosphere Reserve by the United Nations Educational, Scientific, and Cultural Organization (UNESCO) due to its unique ecosystem (Damtie et al., 2017).

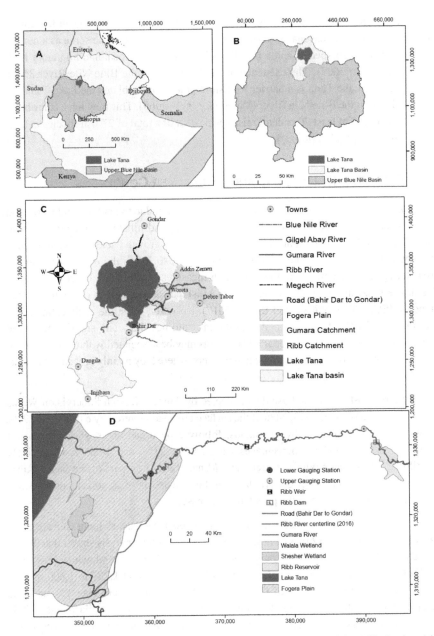

Figure 2.1. Location map of the study area. (A) Ethiopia, (B) Upper Blue Nile Basin with Lake Tana and Lake Tana Basin, (C) Lake Tana Basin including Lake Tana, Fogera Plain, Ribb and Gumara watersheds and (D) The Fogera Plain including the Ribb and Gumara Rivers and the major wetlands.

The lake is surrounded by a low elevation plain which is affected by frequent flooding in the rainy season. The seasonal and perennial wetlands around the lake serve as a nursery site for endemic fish species and a breeding place for several local and migratory birds (Vijverberg et al., 2009). The existence of Tis Issat Fall on the Blue Nile River 35 km downstream of Lake Tana is a barrier for fish migration and isolates the Lake Tana fish community from the Blue Nile River (Vijverberg et al., 2009). This may have contributed to the existence of at least 20 endemic fish species out of a total of 27 fish species in Lake Tana (Vijverberg et al., 2009).

## 2.2 TOPOGRAPHY, LAND USE AND LAND COVER CONDITIONS

The Lake Tana Basin is characterized by land degradation due to land use and land cover (LULC) changes related to deforestation, overgrazing and hillslope cultivation (Abebe and Minale, 2017), which is mainly caused by population growth (Zeleke and Hurni, 2001; Bewket and Sterk, 2005). The increased population also led to the use of less suitable areas for agriculture without soil conservation measures (Minale and Belete, 2017) and this has resulted in increased soil erosion in hillslopes and sediment deposition in flat areas (SMEC, 2008a; Minale and Belete, 2017). However, coastline sedimentation and erosion around Lake Tana are small and the lake did not show any significant natural water level fluctuations (Poppe et al., 2013). This may be associated with the presence of large plains at the lower reaches of the major rivers, where they mainly breach their banks to create recurrent flooding.

Based on the most recent land use data collected by Amhara Design Supervision Works Enterprise (ADSWE), in 2013, 74% of the surface area was covered by cultivations, 21% by water bodies, and 2.8% by grasslands, followed by urban areas, shrub and forest (Figure 2.2A). In 2002 the percentages were 54.5%, 21%, and 10.35%, respectively (WBISPP, 2002). The study by Garede and Minale (2014) on part of the basin (Ribb) indicated an expansion of agricultural lands at the expense of grass, forest, and wetlands and, this may be attributed to the population increase.

The elevation of the Lake Tana Basin varies from 4,000 m a.s.l, near Mount Guna (eastern part of the basin), to 1,787 m a.s.l at Lake Tana. Topographic analysis shows that 39.3% of the watershed has an elevation range between 2,000 to 3,000 m a.s.l, while 33.6% and 26.5% has an elevation ranging from 1,800 to 2,000 m a.s.l and 1,787 to 1,800 m a.s.l, respectively. Moreover, 36.8% of the watershed has a slope of less than 1%, and it is located around Lake Tana, including the Fogera, Dembia, and Kunzila plains. While 43.6% and 18.3% of the watershed have slopes of 1 to 5% and 5 to 15%, respectively (Figure 2.2B).

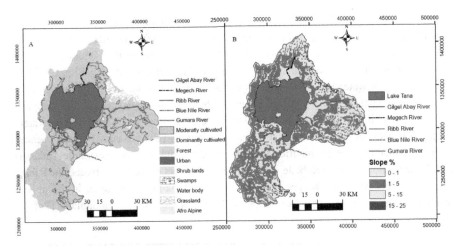

Figure 2.2. Land use (A) and Slope (B) of the Lake Tana Basin with major rivers.

The economy of the basin is dominated by rainfed agriculture. It covers ~80% of the cultivated land and is affected by frequent floods and droughts (Abera, 2017). Flood recession cropping on the floodplains and periphery of Lake Tana and pump irrigation along the major rivers are also practiced by farmers. Cereals, pulses, oilseeds, vegetables, root crops, and fruits are some of the crop categories grown in the basin (Tefera, 2017). Horticulture including flower production along the shore of the lake is recently cultivated by the private investors for export. Nowadays, large, medium, and small scale irrigation projects are also constructed by the government around Lake Tana for its fertile land and available water resources.

## 2.3 GEOLOGY AND SOIL CHARACTERISTICS

The Lake Tana Basin is formed by volcano-tectonic activities and the central part of the basin is dominantly covered with thick alluvial sediments extending below the lake bed surface (Chorowicz et al., 1998; Nigate et al., 2017). The basin has three major aquifers: the Tertiary volcanic, the Quaternary volcanic, and the alluvio lacustrine sediments (Kebede, 2012; Nigate et al., 2016). The Trap Series (Oligocene to Miocene) volcanic flows and edifices, Quaternary basalt and Quaternary sediments (comprises alluvial lake deposits) are the main geological units of the basin and mainly found in the highlands and cliffs, in the lowlands and the floodplain along the lake coastline (Prave et al., 2016; Nigate et al., 2020). The surface area of the basin is dominated by Tarmaber Basalt with intercalated tuffs and basaltic lava flows (SMEC, 2008a). It mainly comprises the granitic and PreCambrian metamorphic basement rocks (SMEC, 2008a). The headwaters of the Lake Tana and its surrounding are mainly dominated by the Tertiary Ashangi Basalt formations. The Fogera and Dembia floodplains (eastern and northern part of the Lake Tana, Figure 2.3A) are dominated by alluvial sediments (ranging from clay to gravel) with an unknown thickness (SMEC, 2008a). The Ribb and Gumara watersheds are dominated by Tertiary volcanic formations, while the Fogera Plain and along the major rivers are dominated by Quaternary lacustrine sediments (Figure 2.3A). The Gilgel Abay watershed (the southern part of the basin) is dominated by Quaternary basalts and volcanic and characterized by the existence of different springs, for example, spring Areke (140 l/s) and Lomi (40 l/s), which are the sources for Bahir Dar city water supply (Kebede, 2012; Nigate et al., 2016). The basin is also characterized by the presence of dikes and faults which are manifested by the existence of hot springs (example Andassa and Wanzaye hot springs) (Kebede, 2012). The headwaters of the major rivers (the Ribb, Gumara, Gilgel Abay and Megech) are characterized mainly by bedrock in the upper reaches and meandering alluvial formations at their lower reaches and floodplains (Poppe et al., 2013).

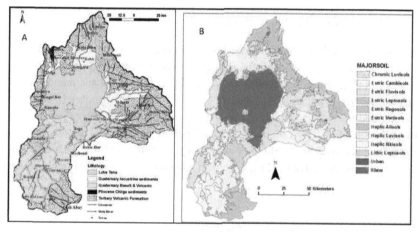

Figure 2.3. The geological formations (A) and the soil distribution (BCEOM, 1998) (B) of the Lake Tana Basin (Source: Nigate (2019)).

The dominant soil source is the weathered basalt profiles (SMEC, 2008a). The major soil types in the basin include Eutric Cambisols, Chromic Luvisols, Eutric Fluvisols, Eutric Regosols, Eutric Leptosols, Eutric Vertisols, Haplic Luvisols, Haplic Alisols, Lithic Leptosols and Haplic Nitisols (BCEOM, 1998; Setegn et al., 2008) (Figure 2.3B). The central and the southern parts of the basin are mainly dominated by shallow Leptosols (Getahun and G. Selassie, 2017). The flat area around the lake is dominated by Vertisols and Fluvisols, while Leptosols dominate the mountainous regions.

## 2.4 RAINFALL AND CLIMATIC CONDITION

The drainage network of the Lake Tana Basin is of dendritic type with less vegetation cover and high rainfall intensity. Based on precipitation, four seasons can be distinguished: the main rainy season (June-August), the post-rainy season (September-November), the dry season (December-February) and the per-rainyseason (March-May) (Negash et al., 2011). The basin experiences spatial and temporal rainfall and temperature variations associated with the movement of Inter-Tropical Convergence Zone (ITCZ) (SMEC, 2008b; Abebe et al., 2017), where trade winds of the Northern and Southern Hemisphere with heat and moisture converge to form a zone of cloudiness and precipitation. During the main rainy season, there is a north-south movement of the ITCZ with moist air coming from the Indian and Atlantic Oceans. In the dry season, the ITCZ moves southwards (Awulachew et al., 2009; Abebe et al., 2017). The mean annual temperature of the basin is 20 °C, characterized by high diurnal and small seasonal variations (Kebede et al., 2011). The maximum and minimum values are observed near the lakeshore and in the Guna and Semine Mountains on the northern and eastern side of the basin (Figure 2.4b), respectively (Kebede et al., 2006; Beyene, 2018). 91% of the basin is characterized by the tepid thermal zone, while 7.8%, 0.64%, and 0.54% of the basin are dominated by cool, warm and cold zones, respectively (Abebe et al., 2017). The tepid thermal zone, found at elevations that are less than 2500 m a.s.l., is the most suitable climatic condition for living, crop and animal production. The northern and eastern parts of the basin have high temperature and low rainfall values, while the southern and western parts have a high rainfall and low temperature values (Figure 2.4, prepared Beyene (2018) using the mean annual rainfall and temperature values from 2000 to 2015).

The rainfall of the Lake Tana Basin has unimodal characteristics with a mean annual value of 1,326 mm: the maximum (1,600 mm) and minimum (1,200 mm) values are observed in the southern and northern part of the watershed (Figure 2.4a), respectively (SMEC, 2008a; Leggesse and Beyene, 2017; Beyene, 2018). Rainfall is characterized by high intensity with erratic nature and varies with altitude, slope and wind direction (SMEC, 2008a; Abebe et al., 2017). Rainfall characteristics affect the discharge volume of the rivers draining to Lake Tana. Typically, 70-90% of the annual rainfall occurs in the main rainy season (June to September) (Uhlenbrook et al., 2010; Tekleab et al., 2013; Weldegerima et al., 2018). In July/August the rainfall reaches its maximum (250-330 mm per month) (Setegn et al., 2008), while in other months there is little rainfall and crop production should be supplemented with irrigation (Beyene, 2018).

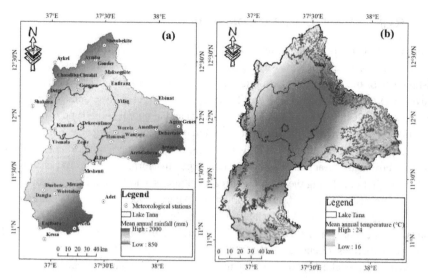

Figure 2.4. The mean annual rainfall including towns (a) and the temperature of the Lake Tana Basin (b) (Source: Beyene (2018)).

## 2.5  WATER RESOURCES DEVELOPMENTS

### 2.5.1 Hydropower Developments

The Tis Abay-I hydropower plant (Figure 2.5) was constructed in 1964 on the Blue Nile River by the then Ethiopian Electricity Light and Power Authority (EELPA) currently named as Ethiopian Electricity Power Corporation (EEPCo) close to the Tis Issat Fall, which is 35 km downstream from the Lake Tana (McCartney et al., 2010). The power plant had an installed capacity of 1.4 MW having a head of 40 m and diverted upstream of Tis Issat waterfall, which is one of the famous tourist attractions (Awulachew et al., 2009; Belete, 2013). Since 1996, the Lake Tana water level is regulated at the outlet of the Blue Nile River by the Chara-Chara weir (Figure 2.5). The purpose of this is to maintain the water level of the lake between 1,784 and 1,787 m a.s.l and to store more than 9,100 Mm³ of water (SMEC, 2008a; McCartney et al., 2010). This resulted in increased dry season discharge and decreased wet season discharge of the Blue Nile River. Before regulation, the lake level variation was influenced by the inflows from the watersheds and used to attain its maximum in September and its minimum in June (McCartney et al., 2010). In 1995, the increased volume storage of the lake convinced the government to construct another power plant (Tis Abay-II) 100 m downstream of Tis Abay-I, which became operational in 2001 with an installed capacity of 72 MW (SMEC, 2008a; McCartney et al., 2010).

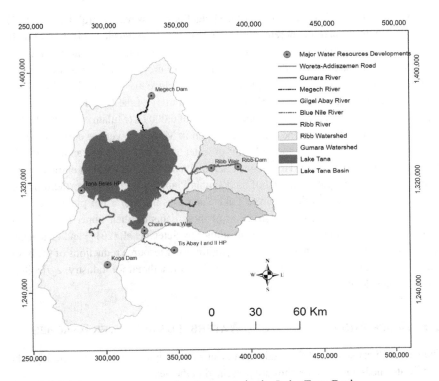

Figure 2.5. Major water resources developments in the Lake Tana Basin.

Another major hydropower plant in the basin is the Tana Beles, located on the western side of the lake (Figure 2.5), which became operational in May 2010. The scheme transfers 160 m³/s of water from the lake to the Beles River to produce 460 MW of power by means of a 12 km long tunnel with a diameter of 7.1 m and a net head of 311 m (SMEC, 2008a). The power plant withdraws an average flow of 77 m³/s from the lake if the power plant is operated with an average plant factor of 48% (SMEC, 2008a; Annys et al., 2019). In addition, part of the water released from the Tana Beles hydropower plant is used to irrigate 75,000 ha of command area in the Upper Beles Basin, mainly for the sugar cane production for the Tana Beles Sugar Factory (Annys et al., 2019).

## 2.5.2 Irrigation Developments

The Lake Tana and the Upper Beles Basins have been identified as one of the five growth corridors by the Ethiopian government for its fertile land and ample availability of water for irrigation (Belete, 2013). The Lake Tana Basin alone has an irrigation potential of ~114,000 ha (SMEC, 2008a). Another study by Worqlul et al. (2015) indicated that around 11% of the basin area (130,000 ha) is suitable for medium scale (between 200 and

3,000 ha) and large scale (>3,000 ha) surface irrigations. However, Worqlul et al. (2015) indicated that only 3% of the potentially irrigable land can be developed if 90% of daily discharges of major rivers are considered indicating the importance of water storage by dams for the dry season. For this, different irrigation projects in the basin are either constructed or under construction or planned. These include the construction of the 21 m high Koga Dam (Figure 2.5) in the Koga River, a tributary of the Gilgel Abay River, to store 83 Mm³ of water and irrigate a command area of 6,000 ha (Mulatu et al., 2012). The Ribb Dam in the Ribb River and the Megech Dam in Megech River are currently under construction to store 234 Mm³ and 181 Mm³ of water to irrigate command areas of 15,000 ha and 7,300 ha, respectively. The Megech Seraba Pump Irrigation project extracts water from Lake Tana to irrigate 4,000 ha of the Dembia Plain, located in the northern part of the lake. The major rivers that drain to Lake Tana pass through a flat plain and cause recurrent flooding. Hence, the construction of these dams on their upper reach is also sought as a way to reduce flooding. Recent studies by Setegn et al. (2011) and Goshu and Aynalem (2017) indicated that there may be future water resource reductions of the basin due to climate changes in the future which may become a threat for industry, agriculture and urban developments.

## 2.6 DESCRIPTION OF THE STUDY WATERSHEDS AND RIVER REACHES

The geographical setting of the study area is shown in Figure 2.6 and the details of the study watersheds and river reaches are presented in the next sections.

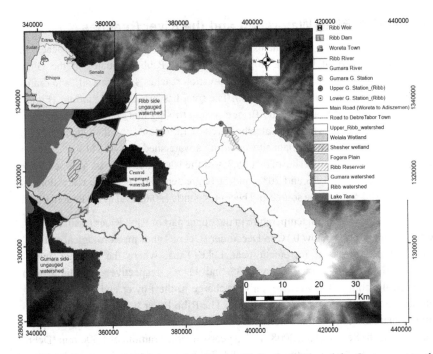

Figure 2.6. Geographical setting of the Fogera Plain, the Ribb and the Gumara watersheds with rivers and wetlands indicated.

In the past the Lake Tana experienced three distinct regulation periods: (1) a rise by 45 cm in 1995-2001, (2) a drop by 53 cm in 2001-2010, and (3) a rise by 62 cm in 2010-2015 (Figure 2.7). Increased flooding of the Fogera Plain and the recent Ribb River channel avulsion in 2008 have been often attributed to past lake level regulation. The water levels of the lake have been monitored since 1959 at Bahir Dar station, and the long-term annual water level variation before regulation was found to be 1.6 m.

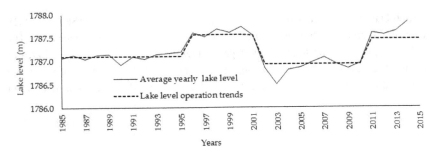

Figure 2.7. Mean annual Lake Tana level since 1985 (data source: Ministry of Water, Irrigation and Energy, Ethiopia).

## 2.6.1 The Ribb River Watershed and the River Reaches

The Ribb River is one of the components of the Upper Blue Nile Basin system located in the North-western part of Ethiopia (Figure 2.1C). The Ribb originates in the Guna Mountains of the Lake Tana Basin, where the elevation reaches 4,000 m a.s.l., and drains to Lake Tana at an elevation of 1,787 m a.s.l. The river has two gauging stations (Figure 2.6): the Lower, located near the Ribb Bridge (road crossing that connects the cities of Bahir Dar and Gondar), covering a watershed area of 1,592 km², and the Upper, located 3.5 km downstream of the Ribb Dam site, covering a watershed area of 844 km². Terrain analysis of the Ribb watershed shows that 18.3% of the watershed area has a slope of less than 10%, 42% between 10% and 20%, and 50% greater than 20%. The river slope is relatively steep (0.3%) near the source and becomes gentle (~0.037%) in the Fogera Plain.

The average annual maximum temperature in the upper part of Ribb River Basin is 27 °C, while the minimum falls below 0 °C in December (Debre Tabor meteorological station). The temperature shows an increment near Lake Tana, where the average annual maximum and minimum values become 35 and 11 °C, respectively (Addis Zemen meteorological station). The average daily discharge at the Lower and Upper gauging stations are 15 m³/s and 8.3 m³/s, respectively. The Ribb River flow depends mainly on the rainfall characteristics of the watershed. The climatic condition of the basin is tropical highland monsoon (Setegn et al., 2008) with a yearly average rainfall of 1,300 mm (Debre Tabor Metrological station of the years 1988–2015); 80% occurs between June and September. The discharge analysis at the Lower gauging station (2000 - 2010) shows that 75%, 10% and 15% of the discharges, which were above the estimated bankfull values, occurred in August, July, and September, respectively. The annual rainfall trend analysis of the watershed, assuming the Debre Tabor meteorological station is representative, does not show any increasing or decreasing trends in the last three decades (Figure 2.8) although there are considerable year-to-year variations. Similar results were also found by Tekleab et al. (2013) for the Upper Blue Nile Basin, Abate et al. (2015) for the Gumara River (adjacent to the Ribb) and, Tesemma et al. (2010) and Hurni et al. (2005) for the entire Blue Nile Basin.

The Ribb Dam is under construction 77 km upstream of the Lake Tana (Section 2.5.2). The location is reachable through a 40 km dry weather road from Addis Zemen town. The dam is located 3.5 km upstream of the Upper gauging station with a watershed area of 715 km². The reservoir will have a capacity to impound 234 million m³ of water and inundate 10 km² of surface area at the Normal Pool Level (NPL) elevation of 1,940 m a.s.l.

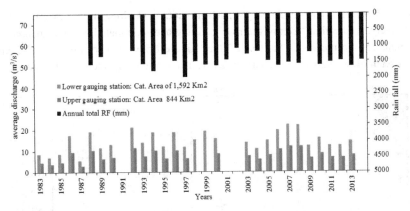

Figure 2.8. Yearly rainfall 1988-2014 and average Ribb River discharges 1983-2013 measured at the Lower and Upper gauging stations. Source: National Meteorological Agency and Ministry of Water, Irrigation and Energy, Ethiopia.

A 64 m long diversion weir is under construction on the main river channel 30 km downstream of the dam location to divert part of the river flow to the command areas (WWDSE and TAHAL, 2007). This means that after the start of dam operation, the river reach between the dam and the weir will convey a strongly regulated water discharge to the weir site, from where an irrigation canal will depart. In the dry season, the flow rate will include the discharge meant for irrigation (16.4 $m^3$/s or larger) plus an environmental flow of 0.17 $m^3$/s or less for the downstream reaches (BRLi and MCE, 2010). The water released from the dam will be relatively clear of sediment and for this reason degradation may be expected in the alluvial parts of the reaches. During low flow conditions, the downstream Ribb River reaches will only receive the environmental flow, which varies between 0.15 $m^3$/s and 0.17 $m^3$/s. This small discharge will nevertheless be larger than the current dry-season flow when the river is regularly dry (Figure 2.14C). The operations of the dam will thus strongly affect both the water discharge regime and the sediment transport in the downstream reaches (e.g., (Williams and Wolman, 1984; Brandt, 2000; Graf, 2006)). In response, the river will adjust its morphology to the new conditions (e.g., (Kondolf, 1997; Khan et al., 2014)).

The study river reach has a length of 77 km, starting at the Ribb Dam site to the Lake Tana. The reach is here sub-divided in four parts: the Upper-I, the Upper-II, the Middle and the Lower reaches (Figure 2.9). The presence of river bed fixation (rock outcrops and weir), the junction of major tributaries, the longitudinal slope, the channel width, and the presence of anthropogenic interventions (sand mining, embankment construction and water extraction) are the basis of this sub-division.

The Upper-I river reach comprises the first 9.7 km downstream of the dam under-construction. In this reach, the river passes through a moderately deep gorge where the banks rise steeply. The reach-averaged river width and slope are 65 m and 0.3%, respectively. In the first 8.5 km, the river bed is gravel-dominated, whereas numerous rock outcrops are present in the last 1.2 km. An old bridge crosses the river 4.25 km downstream of the dam site, whereas currently, a new bridge is under-construction.

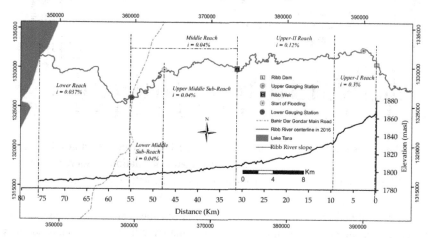

Figure 2.9. Ribb River study reach, sub-reaches and river bed profile from the dam site to Lake Tana. The river alignment is derived from the Google Earth images of 2016. The slope is obtained from the 30 m resolution Digital Elevation Models (DEM) of Advanced Spaceborne Thermal Emission and from the Reflection Radiometer (ASTER) satellite images of 2014.

The Upper-II Reach comprises 22.3 km of Ribb River, from the end of the rock escarpment to the weir under-construction, and shows a much milder slope of 0.12%. The river channel is here characterized by a meandering planform with little bank erosion and accretion. The alluvial river bed is dominated by a mixture of sand and gravel. In particular, gravel dominates the outer side of sharp bends. Further downstream, the gravel component reduces and the river bed becomes sand-dominated. The river banks in the Upper-II reach are dominated by compacted clay and sand formations and show hard rocks at some locations.

The Middle Reach (sub-divided in Upper Middle and Lower Middle) is 25 km long, from the weir site to the Ribb Bridge (Figure 2.9). The reach is sub-divided based on the existence of embankments for flood protection. The Upper Middle sub-reach is 15 km long and extends from the weir site to the start of the embankment, while the Lower Middle sub-reach forms the remaining 10 km where embankments are present. This part of the reach experiences flooding due to reduced channel conveyance capacity (SMEC, 2008a). The Middle reaches have a relatively gentle slope (0.04%) and a sandy bed.

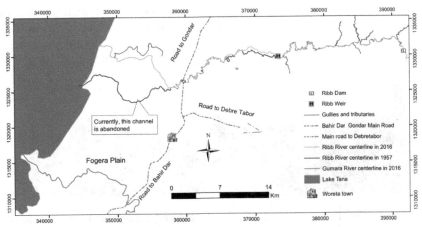

Figure 2.10. Ribb River alignments in 1957 and 2016 including gullies and tributaries. The area between the Lake Tana and the road to Gondar and Bahir Dar is part of the Fogera Plain, which is shared by the Ribb and the Gumara rivers.

The Lower Reach is 20 km long, from the Ribb Bridge to the Lake Tana. In the past, the river channel bifurcates 4 km downstream of the bridge (Figure 2.10). Since 2008, the old channel, to the left of the current one, is completely silted up and serves as a dry-weather road. During peak flows, however, the old channel still conveys a small part of the flow to the lake. Further downstream, the river loses its channel for a distance of 3.5 km (i.e. from 70.2 km to 73.7 km from the Ribb Dam). Here, high-flow water spreads over the floodplain until it finds a well-defined channel again near a locality called Boled Gote. Farmers plant maize in the shallow river channel during the dry season, taking advantage of residual moisture (Figure 2.11). The river banks in the Middle and Lower reaches are made of compacted clay and silt materials.

Figure 2.11. Lower Ribb River reach. (A) The point where the river loses its channel. (B) Maize plantation using residual moisture along the river channel downstream of location A.

21

The Ribb River channel has a meandering planform with sporadic central bars in the Upper-I reach where the channel is wider. Point bars inside river bends and alternate bars are observed in the Upper-II, Middle and Lower River reach. Figure 2.12 shows the river bed and bank characteristics in the Upper-I reach and the Lower Middle sub-reach.

Figure 2.12. River bed and bank material in the Upper-I (A) and the Lower Middle (B) River reaches (March, 2016).

The granulometric curves of the bed material (Figure 2.13) show that the Ribb River sediment becomes finer in downstream direction, as expected (Williams and Wolman, 1984; Kinzel and Runge, 2010). The river bed near the dam site is dominated by a mixture of gravel (65%) and sand (35%). The bed material gradually changes to sand near the weir site (60% sand and 40% gravel). Further downstream, the river bed is composed of 90% sand. The median grain size, $D_{50}$, is 7 mm in the Upper I and Upper II reaches and decreases to 0.65 mm in the Middle (the Upper and the Lower) reaches and 0.35 mm in the Lower reaches. Historical data on sediment grain size are lacking and this does not allow assessing their temporal evolution.

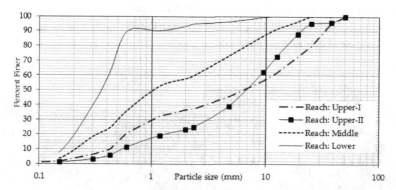

Figure 2.13. Grain size distribution of Ribb River bed material downstream of the Ribb Dam site, upstream and downstream of the weir site and upstream of the Ribb Bridge, at UTM coordinates 37P 0388475, 1332080, 1872 m a.s.l; 37P 0373692, 1329576, 1819 m a.s.l; 37P 0368803, 1330762, 1804 m a.s.l and 37P 0359808, 1326196, 1795 m a.s.l, respectively (samples collected in March, 2016).

The Ribb River is further affected by extensive water withdrawal for irrigation during the dry season using diesel pumps (Figure 2.14A). Farmers also excavate trenches over the embankment to reduce the length of their pump hose and the suction head to extract water for dry time irrigation, which may cause breaching during high flows. However, drainage water from irrigation may return to the river system as seepage in the soil, thereby wetting up the streambank materials. This could contribute to the destabilization of streambanks as shown in Figure 2.14B, where a block of soil is about to fall in the river. Failed material accumulates at the bank toe. This material may be either transported downstream during high flows or become stabilized by the growth of vegetation during low flows. Intensive water pumping and a prolonged dry season occasionally result in complete drying of the river in the Lower reach (Figure 2.14C). Moreover, the discharge reduction may affect the sediment transport capacity of the river system, resulting in sedimentation. In the long term, water withdrawal from the river leads to increased longitudinal bed slope and reduced water depth (Jansen et al., 1979).

Figure 2.14. (A) Water extraction with a diesel pump for irrigation, (B) Irrigation canal to transport extracted water, which has been breached, and adjacent failing banks and (C) Ribb River during the dry season. The farmers excavate holes in the sandy river bed to get water for cattle and backyard irrigation. The photos were taken during a field campaign in April 2017.

The river system experiences also significant sediment extraction from the main channel bed (Figure 2.15A). The prolonged sand mining activity in the Middle and Lower River reaches has clearly affected the natural river bed topography, altering the shape and height of sand bars, as well as the sediment balance of the river. These activities have steadily increased in the last 30 years.

Upstream of the Lower gauging station, embankments are built on bank tops without any setback distance, which increases the risk of bank failure, although there is an attempt to protect river banks at bends with gabions. Additionally, channel enlargement and deepening have taken place at shallow and narrow sections to increase the flow conveyance (Figure 2.15B). To facilitate the release of floodwater to farmland for agricultural use, outlet structures have been constructed as low height weirs controlled by gates at selected locations (Figure 2.15C). However, the release of high velocity floodwater may create deep gullies unless corrective measures are taken.

Figure 2.15. (A) Excavated sand accumulation by sand miners in the Middle River Reach. (B) River channel enlargement and deepening to increase flow conveyance. Excavated material is used further downstream for embankment construction. (C) Flood outlet structure to the farmlands as low height weir (April, 2016).

## 2.6.2 The Gumara River Watershed

The Gumara River has one gauging station near the Gumara Bridge, the road connecting Bahir Dar and Gondar towns, with a watershed area of 1,412 km$^2$ (Figure 2.6). The watershed is characterized by an undulating topography with an elevation that varies from 3,704 m a.s.l. at the source to 1,787 m a.s.l. near the Lake Tana. 25.7% of the watershed area, mainly found in the Fogera Plain, has a slope that is less than 10%, while 33.2% of the area has a slope between 10% and 20%. The remaining 41.1% is characterized by steep slopes, i.e. greater than 20%. Like the Ribb River, the reach of the Gumara River downstream of the bridge is affected by recurrent flooding (Abate et al., 2015). Like other rivers in the Lake Tana Basin, the sediment concentration of the river becomes maximum at the beginning of the main rainy season (June and July) when the ground surface is not covered by vegetation and the soil integrity is affected by ploughing (Easton et al., 2010; Abate, 2016).

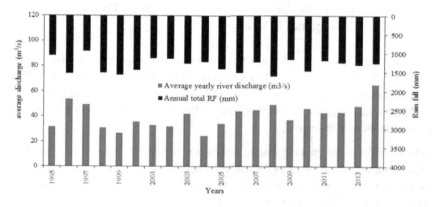

Figure 2.16. Yearly rainfall from the year 1995 to 2014 of Woreta Meteorological station and the average Gumara River discharges from the year 1983 to 2013. Data source: National Meteorological Agency and Ministry of Water, Irrigation and Energy, Ethiopia.

The bankfull discharge of the Gumara River, estimated using the Flow Duration Curve method, is 240 $m^3/s$ and 254 $m^3/s$ for a return period of 1.5 and 2 years, respectively. The discharge analysis of the river at the gauging station for the period 2000-2015 shows that 70% of the discharges are above the estimated bankfull values are occurred in August while 15% in July and 15% in September. The average yearly watershed yields of the Gumara and Ribb, estimated using the measured time-series discharge data at the Lower gauging station, are 3.8 $m^3/s/km^2$ and 10.42 $m^3/s/km^2$ for the Ribb and the Gumara watersheds, respectively. The significantly differing values for these adjacent rivers may be attributed to the location of the gauging stations (even if they are both located in the floodplain where the peak discharges overtop the river banks upstream of the stations), moderate differences in land use/land cover and major variations in soil types. Moreover, the trend analysis of the Gumara river discharge and the rainfall at Woreta meteorological station (assumed representative for the watershed) (Figure 2.16) does not show any increasing or decreasing trends, as described by Abate et al. (2015).

## 2.6.3 The Fogera Plain

The Fogera Plain (Figure 2.6) experiences recurrent flooding, mainly in August and in the beginning of September (SMEC, 2008b; Mulatu et al., 2018). The flooding events of 2006 and 2010 were exceptional for their long durations and the high number of fatalities and evacuated people. Flooding affected farmlands, houses, health centers, water facilities and other infrastructure (ENTRO, 2010). In 2006, more than 30,000 persons were evacuated, 45 persons died, a thousand hectares of agricultural lands were damaged, and infrastructure was demolished (ENTRO, 2010). The average flooded area of the Fogera Plain was estimated by SMEC (2008b) in the period 2001-2006 and found to be 275 $km^2$ in July. The inundation area shows an increasing trend in recent years (ENTRO, 2010) even though the rainfall pattern in the watersheds does not show any increasing or decreasing trends (Figure 2.8 and Figure 2.16). This may be caused by land use and land cover changes due to increased population, embankment construction and sedimentation in the river channels (SMEC, 2008b; Mulatu et al., 2018). Flooding of the Fogera Plain is mainly caused by (i) overbank flow from the Ribb and Gumara Rivers (SMEC, 2008b; Abate et al., 2015; Mulatu et al., 2018); (ii) direct rainfall on the clay soil of the plain in combination with low gradients and poor drainage to the main river systems and local runoff generated by gullies that drain into depressions (Liu et al., 2008; SMEC, 2008b); (iii) backwater from Lake Tana due to lake level regulation at its outlet (SMEC, 2008b) and (iv) additional water flow to the plain from ungauged watersheds surrounding the Ribb and Gumara Rivers, which are either drained by gullies or by overland flows (Figure 2.6). The regional government is currently building low-height embankments along the Ribb River in an attempt to temporarily reduce flooding. A 15 km long embankment was constructed in 2016 (8 km upstream of the bridge and 7 km downstream of the bridge).

Of the 383 km$^2$ of the Fogera Plain (Figure 2.6), 288 km$^2$ are dominated by cultivated land, followed by 56.2 km$^2$ of water bodies and wetlands, 31.1 km$^2$ of scattered farm villages, and the remaining 7.3 km$^2$ are covered by shrub, bushlands, forest, and grass. The major soil types of the floodplain include Chromic Luvisols, Eutric Leptosols, Eutric Vertisols, and Umbria Andosols. Moreover, the soil characteristic coverage analysis shows that 91% of the plain is dominated by clay with low infiltration rates. Two major wetlands are present in the plain, namely "Welala" and "Shesher" (Figure 2.17). The dominant soil type of the wetlands is the gleysol, characterized by the prolonged saturation and low rate of infiltration.

The wetlands have a direct connection with the rivers that frequently supply water and nutrients, as they are located in the active flooding zone. They are the spawning and breeding grounds of locally endangered fish species (Francis and Aynalem, 2007; Mohammed and Mengist, 2019), of which the Clarias gariepinus and the Labeobarbus are the dominant ones (Anteneh et al., 2012). These fish species migrate upriver during the rainy season (July to October) to spawn in the well-oxygenated, clear and fast-flowing water and return to the Lake Tana when the flooding recedes in October to December (Anteneh et al., 2012; Abebe et al., 2020). The wetlands are also an attractive area for birds, feeding on the remnants of dead fish, insects in various types of micro-habitats, cereals, vegetables and, fruits (Mitsch, 2005; Aynalem, 2017). They provide shelter and roosting places for globally endangered and endemic birds (Mundt, 2011; Negash et al., 2011) and are an important stopover and breeding place for migratory birds, such as the Egyptian Goose, the Curlew Sandpiper, the Common Crane, the Black Crowned Crane, the Spur-winged Goose, the Ruf and other geese (Francis and Aynalem, 2007; Negash et al., 2011; Aynalem, 2017). The benthic macro-invertebrates in the wetlands were studied by Negash et al. (2011), who found that Chironomidae (70%) and Oligochaeta (25%) are the most abundant.

The wetlands are a resource for several economic activities, like cattle watering and grazing, fishing, farming, and sand mining (Negash et al., 2011; Wondie, 2018). Water extraction for irrigation and conversion of wetlands to agricultural lands in the dry season have affected the quantity of water and the biodiversity of the wetlands (Wondie, 2018; Mohammed and Mengist, 2019). Studies by Wondie (2018) and Mohammed and Mengist (2019) indicate that the wetland surface area decreases at an alarming rate due to intensive agriculture, free grazing, and population density. Moreover, the land near the Lake Tana and around the wetlands has suffered also due to the expansion of water hyacinth (Dersseh et al., 2019). The villagers consider the wetlands as breeding places of common diseases such as mosquitoes (Mohammed and Mengist, 2019).

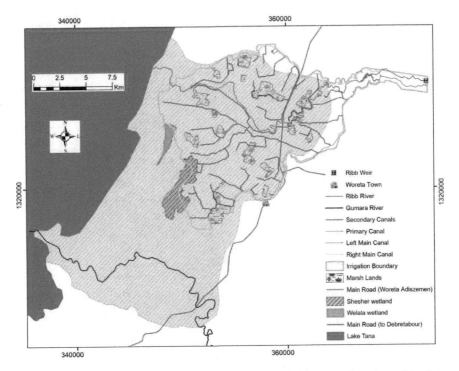

Figure 2.17. Ribb irrigation area with primary, main, secondary canals and marshlands to
be used as a collection chamber for irrigation drainage water (Source: AutoCAD
drawing of the feasibility and detail study of Ribb Dam Irrigation system (2010)).

The implementation of the Ribb Irrigation system will cover a gross area of 16,700 ha of
which 85.7% belongs to the Fogera Plain along the Ribb River (Figure 2.17). The
wetlands (including the Welala and Shesher) and the marsh lands of the floodplain (Figure
2.17) are supposed to serve as a collection chamber/detention basin for the excess
irrigation water from the farmlands in the dry season.

# 3

# MORPHODYNAMIC TRENDS OF THE RIBB RIVER PRIOR TO DAM CONSTRUCTION[1]

This chapter presents the analysis of the current and past Ribb River dynamics based on the newly collected data, aerial photographs, SPOT and Google Earth satellite images. At present, a dam and a weir are under construction to store and divert water for irrigation. This will strongly alter both water and sediment discharges to the downstream river, causing morphological changes. Assessing the current morphodynamic trends is the first necessary step to study the future effects and find ways to mitigate them.

---

[1] This chapter is based on:

Mulatu, C. A., Crosato, A., Moges, M. M., Langendoen, E. J., & McClain, M. (2018). Morphodynamic Trends of the Ribb River, Ethiopia, Prior to Dam Construction. *Geosciences, 8*, 255. doi:10.3390/geosciences8070255.

## 3.1 INTRODUCTION

Alluvial rivers adjust their slope, planform and bed topography in response to sediment and/or water input changes due to either anthropogenic influences, natural events (landslides, etc.), or climate change (Jansen et al., 1979; Latapie et al., 2014). The study of the nature, rate and causes of river channel changes has a particular relevance to areas where high levels of natural and human disturbance threaten engineering structures and property (Gilvear, 1999) as well as the biodiversity of vegetation communities within the river corridor (Gilvear, 1993; Marston et al., 1995; Bravard et al., 1999). Knowledge of morphodynamic trends and their alterations is thus key to proper management of river resources (Gilvear and Winterbottom, 1992).

Assessing the future impact of planned works requires an exhaustive description of the river hydro-morphological characteristics and dynamics prior to the intervention (Winterbottom, 2000). This is necessary, for instance, for the simulation of the hydro-morphological adaptation of the river with a numerical model (e.g., (Nelson et al., 2016)) because it establishes the starting conditions. It is also necessary for the analysis of future field data to unravel the role of the intervention because it offers a base-case scenario for comparison. Knowledge of the current morphology and dynamics is also a key to identifying the effects of on-going climate changes on the functioning of the river and its floodplain. Finally, acquiring this knowledge is the first step for the definition of mitigation strategies.

The main objective of this study is to define the pre-dam situation of the Ribb River system and identify the current morphodynamic trends in order to allow for the assessment of the impact of the planned structures and operations on the downstream reaches. The study includes the analysis of the effects of current Lake Tana level regulation, river embankment for flood control, sand mining and water extraction for irrigation on past and present river bed topography and planform changes.

The methodology includes the analysis of historical data, namely aerial photos, SPOT satellite images, and Google Earth imagery to study the river planimetric changes. The effects of Lake Tana regulation on the Ribb River bed level adjustment propagation in the upstream direction was analyzed considering the morphological time-scale derived by De Vries (1975). The effects of sand mining and water extraction on river bed slope were analyzed using the morphodynamic equilibrium theory developed by Jansen et al. (1979). The work includes the analysis of data provided by the Ministry of Water, Irrigation and Energy, Ethiopia, as well as data from the feasibility study and design documents of the Ribb Irrigation project (WWDSE and TAHAL, 2007), and new field data collected in the framework of this study during two recent field survey campaigns.

## 3.2 DESCRIPTION OF STUDY RIVER REACHES

The detailed description of the Ribb River watershed and the reaches identified for the study, including the anthropogenic aspects and granulometric data are presented in Section 2.6.1.

## 3.3 MATERIALS AND METHODS

This study aims at describing the current state and morphodynamic trends of the Ribb River. It includes data collection and analysis from the literature and two field campaigns (Table 3.1).

Table 3.1. Collected data and their source.

| Data Type | Data Period | Resolution | Source |
|---|---|---|---|
| Black and white aerial photographs | 1957 (Nov. and Dec.) | ~1:55,000 | Ethiopian Mapping Agency |
| | Feb. 1980 | ~1:45,000 | |
| SPOT satellite images | Jan., 2006 | 2.5 m by 2.5 m | Airbus Defence |
| | Nov., 2012 | 2.5 m by 2.5 m | and Space |
| ASTER DEM | Sep., 2014 | 30 m by 30 m | https://earthexplorer.usgs.gov/ |
| Google Earth satellite images | 2016 | | Maps.google.be |
| River cross-sectional survey | Mar., 2016 | | Field campaign |
| River flow discharge at the Lower gauging station | 1964–2014 | Daily | Ministry of Water, Irrigation and Energy |
| River flow discharge at the Upper gauging station | 1980–2013 | Daily | Ministry of Water, Irrigation and Energy |
| Water level at the Lower gauging station | 1980–2010 | Daily | Ministry of Water, Irrigation and Energy |
| Lake Tana level at Bahir Dar station | 1960–2015 | Daily | Ministry of Water, Irrigation and Energy |
| River bed-material samples | Mar., 2016 | | Field campaign |
| Rainfall at Debre Tabor Meteorological station | 1988–2015 | Daily | National Meteorological Agency, Ethiopia |

The analysis includes:

    i.    the description of the river discharge regime,

    ii.    the assessment of bed level changes and sediment transport rates,

    iii.    the description of the historical channel evolution (alignment, width),

    iv.    the assessment of the past adjustment of river slope to sediment mining and water withdrawal, and

    v.    the assessment of propagation time of river bed level adjustment to Lake Tana water level regulation.

The methods adopted are described below.

## 3.3.1 River Flow Regime

The characterization of the river discharge regime at the gauging station locations (the Lower and the Upper) is based on the Flow Duration Curves (FDCs) using the daily time-series of measured discharge covering the period 1983–2010.

The analysis includes the assessment of the bankfull discharge, an important reference condition for river flow and morphology (Parker et al., 2007; Wilkerson and Parker, 2010), which can be derived following different approaches like direct measurements (Leopold and Wolman, 1957; Van den Berg, 1995; Crosato, 2008; Blom et al., 2017), stage-discharge rating curves, and flow frequency analysis of annual maximum series (Leopold et al., 1964; Williams, 1978), even though the practical estimation of flow frequency and magnitude remains difficult (Williams, 1978).

Due to lack of measured data on river channel geometry and water levels along the river course, a flow frequency analysis based on Gumbel extreme distribution was applied, assuming that the annual peak discharge is equal to the instantaneous peak discharge value of each year and that the discharges having return period of 1.5 and 2.0 years are representative for the bankfull conditions (Leopold et al., 1964; Shaw et al., 2010; Vargas-Luna et al., 2018). The maximum daily discharge of each year is filtered from the daily discharge time-series. The discharge magnitude ($Q_T$) having a return period of $T_r$ years is then given by:

$$Q_T = Q_{avg} + K_T . \sigma \tag{3.1}$$

in which $Q_{avg}$ is the average value of the annual peak flows (m$^3$/s), $\sigma$ is the standard deviation of the flows (m$^3$/s), and $K_T$ is the dimensionless frequency factor given by:

$$K_T = \frac{\sqrt{6}}{\pi} \left\{ \lambda + \ln \left[ \ln \left( \frac{T_r}{T_r - 1} \right) \right] \right\} \tag{3.2}$$

where λ is the Euler constant (= 0.5772).

The bankfull discharge is also estimated as a geometrical bankfull condition (Vargas-Luna et al., 2018) for the locations where the cross-sections were measured during the two field campaigns, and then compared to the values obtained by applying the Gumbel method.

### 3.3.2 Bed Level Changes and Sediment Transport Rates

River bed level changes were assessed near the Ribb Bridge based on the analysis of the temporal evolution of the relation between discharges and water levels ('stage-discharge relationship') at the Lower hydrometric station.

The sediment transport rates of the Ribb River were estimated using the most widely used sediment transport formulae. The Meyer-Peter and Müller (1948) formula, revised by Wong and Parker (2006), was applied to the Upper-I and II reaches, both dominated by gravel (Equation (3.3)), while the Engelund and Hansen (1967) formula, was applied to the Middle (the Upper and the Lower) sub-reaches, dominated by sand (Equation (3.4)):

$$Q_s = \frac{4B\sqrt{g}}{\Delta}\left(hi - 0.047\Delta D_m\right)^{\frac{3}{2}} \tag{3.3}$$

$$Q_s = \frac{Bu^5}{20C^2\Delta^2 D_{50}\sqrt{g}} \tag{3.4}$$

in which $Q_s$ is the volumetric sediment transport rate without pores (m³/s), B and h are the reach-averaged channel width and flow depth (m), respectively, i is the water surface slope (-), u is the flow velocity (m/s), C is the Chézy coefficient (m$^{1/2}$/s), $\Delta$ is the submerged specific gravity of sediment (1.65), $D_{50}$ is the median grain size (m), $D_m$ arithmetic mean grain size (m) given by $D_m = \Sigma p_i D_i$, $p_i$ is the probability of the size fraction with diameter $D_i$, (m) and g is acceleration due to gravity (9.81 m/s²).

The sediment transport capacity in the Upper-I reach and in the Lower Middle sub-reach was computed using the daily river discharges measured at the Upper and Lower gauging stations, respectively, since these stations are located within the reaches. For the Upper-II reach and the Upper Middle sub-reach, the daily discharges were derived based on watershed proportion. This was done by assuming that the watershed runoff yield is not influenced by watershed properties and rainfall distribution. The Lower gauging station is located at the start of the Lower reach where a large part of the peak discharges flows on the floodplains and through the remains of the silted-up channel (old river course) (SMEC, 2008b; Dessie et al., 2014). This means that the discharge measured at the Lower gauging station is not representative of the water flowing through the (new) main channel of the river in its lowest course near Lake Tana.

In 1959, the U.S. Bureau of Reclamation installed an automatic water level recorder in a cylindrical housing at a location just upstream of the Ribb River Bridge. The height of the cylindrical housing above the floodplain was used here to roughly assess the near-channel floodplain level rise between 1959 and 2016 (Section 3.4.2).

### 3.3.3 Historical Channel Evolution

The historical river alignments were derived from the aerial photographs of 1957 and 1980, the SPOT satellite images of 2006 and 2016, and a Google Earth image of 2016, complemented with a field reconnaissance. The aerial photographs were scanned with 600 dpi (dots per inch) of geometric resolution and 24 bit radiometric resolution of uncompressed gray scale, whereas the camera calibration reports (fiducial marks and focal lengths) are used to adjust the interior and exterior orientation of the photographs. The Ground Control Points (GCPs) were established during the field campaigns at fixed locations, like road crossings, monuments, and bridges, and some additional GCPs were identified from the SPOT satellite image of 2012 (the x and y values) were used for ortho-rectification with ENVI 4.3. Topographical elevations were derived from the 30 m resolution digital elevation model (ASTER DEM of the year 2014). The ortho-rectified images were then mosaicked to create one image covering the study river reaches. ArcGIS 10.3.1 was finally used to digitize the river centerlines and visualize the super-imposed channel alignments. To capture the river centerline well, both aerial photographs and satellite images were digitized for a scale of 1:4000.

The evolution of channel sinuosity, derived from the historical channel alignments, was here used to express the temporal changes of river meandering intensity. The evolution of channel width was determined from the satellite images and from the field data collected during the river cross-sectional survey of March 2016. The current valley length and longitudinal bed slope of the river were derived from the 30 m resolution ASTER Digital Elevation Model of the year 2014.

### 3.3.4 River Slope Adjustment to Interventions

The Equilibrium Theory developed by Jansen et al. (1979) compares two reach-scale morphodynamic equilibrium conditions, one before and the other after one or more interventions, focusing on longitudinal bed slope and reference water depth by combining the following equations:

1. Reach-scale continuity equation of water:

$$Q_w = Bhu \tag{3.5}$$

2. Momentum equation for water, reduced to Chézy's equation for steady uniform flow, simplified for large width to depth ratios:

$$u = C\sqrt{hi} \tag{3.6}$$

3. Simplified sediment transport capacity formula expressed as a power law of flow velocity:

$$q_s = a(u - u_c)^b \tag{3.7}$$

4. Sediment balance equation:

$$Q_s = yBq_s \tag{3.8}$$

where $Q_w$ is the water discharge (m$^3$/s); B, h and u are the reach-averaged channel width (m), flow depth (m) and velocity (m/s), respectively; i is the water level slope (-), which is equal to the bed level slope (equilibrium conditions); C is the Chézy coefficient (m$^{1/2}$/s); $u_c$ is the critical flow velocity for initiation of sediment motion, which can be considered to approach zero in sand bed rivers (m/s); a is the sediment transport proportionality coefficient (-); $Q_s$ is the average annual sediment transport rate (m$^3$/year); y is the proportionality constant to convert seconds to year, and b is the degree of non-linearity of the sediment transport formula (-). The value of b is equal to 5 when using the Engelund and Hansen sediment transport formula (Engelund and Hansen, 1967), Equation (3.7), however, in general larger than 3 and reaches the value of 10 or more in gravel-bed rivers.

The theory was used here to analyze the longitudinal channel slope evolution of the Ribb River since 1980. The Equilibrium Theory was applied to this part of the river to study the effects of those interventions on river bed slope. It is assumed that: (1) in 1980 the river was in natural morphodynamic equilibrium (absence of relevant interventions); (2) the river hydrology has not changed in the last three decades (Figure 2.8); and (3) the river had reached a new morphodynamic equilibrium state by 2014, after more than 30 years of water withdrawal and sediment extraction. The assumption regarding the absence of relevant interventions was supported by the analysis of the historical satellite images (see Section 3.4.3). Two scenarios were considered: (1) 100% of water is extracted during each dry period (December to May), and (2) 100% of water is extracted during each dry period plus a volume of sand of about 6000 m$^3$ is mined every year. These scenarios are realistic, based on recent field observations (see Section 3.4.2). The values of the percentages and the results are found in Section 3.4.4.

Combining the equations introduced in the previous paragraph, the longitudinal bed and water surface slope at the initial (before the intervention) morphodynamic equilibrium is given by:

$$i_0 = \frac{Q_{s0}^{3/b} B^{(1-3/b)}}{a^{3/b} C^2} \left( \frac{1}{\frac{1}{365} \sum_{k=1}^{365} Q_{w0k}^{b/3}} \right)^{3/b} \tag{3.9}$$

in which $i_0$ is the initial reach-averaged longitudinal bed slope (-), $Q_{s0}$ is the initial annual sediment transport rate (m³/year), and $Q_{w0k}$ is the initial average daily river discharge of k days (m³/s).

Equation (3.9) is valid for all equilibrium states. The subscript "0" denotes the natural equilibrium in the year 1980, just before starting the water extraction and sediment mining interventions, and with subscript "∞" the new equilibrium in 2014, after about 34 years of water and sediment extractions. Assuming constant river width, Chézy coefficient and sediment transport parameters (a and b), the ratio between the equilibrium bed slopes in 2014 and 1980 is given by:

$$\frac{i_\infty}{i_0} = \left(\frac{Q_{s\infty}}{Q_{s0}}\right)^{3/b} \left(\frac{\sum_{k=1}^{365} Q_{w0k}^{b/3}}{\sum_{k=1}^{365} Q_{w\infty k}^{b/3}}\right)^{3/b} \tag{3.10}$$

The method is strictly valid for sand-bed rivers only, since it does not take into account the limitation to bed erosion provided by bed armoring. Consequently, for gravel-bed rivers, the method can be applied only in case of sediment deposition (slope increase). It can also be applied to only qualitatively assess the temporal morphological trends in the river (increase by upstream sedimentation or decrease by upstream erosion).

The use of this simplified approach to assess the historical river slope was justified by its successful implementation in recent works. Its predictability was tested by Duró et al. (2016) who compared the slope derived using Equation (3.10) to the results of a 2D morphodynamic model in case of river width changes and found surprisingly strong agreements. Khan et al. (2014) applied the theory to the sand-bed Middle Zambezi River concluding that the historical slope was similar to the present one.

## 3.3.5 Assessment of Propagation of Bed Level Adjustment to Lake Tana Regulation

The past rise of Lake Tana level might have affected the closure of the old Ribb River channel at the bifurcation, since the long-term morphological effect of a persistent downstream rise of the water level is an equal rise of the river bed level. The bed level adjustment propagates in the upstream direction at a speed that depends on the sediment transport rate and on the geometry of the channel. Considering a certain time interval, it is possible to estimate a reference upstream limit of this propagation using the formula derived by De Vries (1975):

$$L = \sqrt{\frac{TbQ_s}{3Bi(1-p)}} \tag{3.11}$$

where L is the distance (m) from the downstream boundary reached by the bed level adjustment in the time interval T (years), and p is the porosity of sediment deposits (assumed to be 0.4 for uniform sand (Frings et al., 2011)).

Equation (3.11) was derived for sand-bed rivers, assuming $u_c = 0$ in Equation (3.7). This assumption is not valid for gravel-bed rivers, where the existence of a clear threshold for sediment motion limits bed erosion. However, the formula can be applied in case of sedimentation, which is the case of the Ribb River adjustment of Lake Tana level rise.

L represents the theoretical distance reached by a change in bed level that corresponds to 50% of the total change that can be expected. For the Ribb River, the downstream boundary corresponds to Lake Tana water level. To assess whether Lake Tana regulation has indeed contributed to the avulsion event, the considered time interval T is the duration of the period between the first rise of lake level (1995) and the year of avulsion (2008): 13 years. Considering that, in 2001, the lake level was lowered again, it is also important to establish whether this has resulted in river bed level lowering at the location and time of the avulsion. For this, the considered time interval is 7 years.

The average annual sediment transport capacity ($Q_s$) of the old Lower River channel is computed adopting the Engelund and Hansen (1967) formula, for which b = 5 (Equation (3.4)), using the historical discharge time-series at the Lower gauging station of the periods 1995–2001 and 2002–2008 (Figure 2.7). Two conditions were considered. In the first one, the old river channel was assumed to be big enough to convey all discharges, including those above the value that represents the current bankfull condition. In the second one, the old channel was assumed to convey all discharges up to the bankfull value, the water flow in the channel being the bankfull flow rate for higher discharges. Considering the first condition allows assessing the upper limit of L. The second condition is more realistic, but still overestimating L since it does not take into account the gradual reduction of channel cross-section that preceded the avulsion event. As historical channel geometry for this reach is lacking, therefore here the current river channel characteristics at the Lower gauging station was used. This may add errors in the assessment, which, considering all the uncertainties can only result in a rough estimate.

## 3.4 RESULTS

### 3.4.1 River Discharge Characterization

The FDCs derived for the gauging stations are shown in Figure 3.1. It is evident that the peak discharge at the Upper gauging station is substantially greater than the one at the Lower gauging station. This is likely due to overbank flow reducing the discharge exceeding the bankfull by as much as 71% at the Lower gauging station (Dessie et al., 2014).

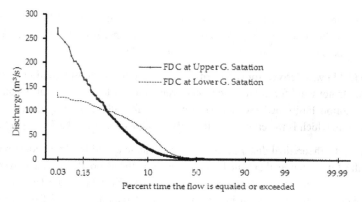

Figure 3.1.Flow duration curves at the Lower and Upper gauging stations (data source: Ministry of Water, Irrigation and Energy).

The values of the discharges having return periods of 1.5 and 2.0 years, here assumed to be representative of the bankfull conditions, are estimated to be 125 m³/s and 196 m³/s for the Upper-I reach, and 117 m³/s and 128 m³/s for the Lower Middle sub-reach, respectively. The values of the bankfull discharge were also computed using Chézy's equation, thus assuming uniform flow, for the measured river cross-sections near the Upper and the Lower gauging stations. Based on expert opinion, a Chézy coefficient of 35 m^{1/2}/s can be considered appropriate for the river. The values obtained using the current longitudinal slope (Figure 2.9) are 150 m³/s and 110 m³/s for the Upper and the Lower gauging stations, respectively. These values are close to the values obtained using the statistical method. For this reason, it can be concluded that the bankfull discharge in the river reach immediately downstream of the Ribb Dam site has the order of magnitude of 150 m³/s and in the reach immediately upstream of the Ribb Bridge it has order of magnitude of 110 m³/s.

## 3.4.2 Bed Level Changes and Sediment Transport Rates

The time-series of daily water levels at the Lower gauging station shows that the river bed rose by 2.2 m (14.7 cm/year) and 0.7 m (4.4 cm/year) between 1980 and 1995 and between 1995 and 2010, respectively, which is 2.9 m in 30 years (1980 to 2010) (Figure 3.2). Similarly, the nearby Gumara River bed rose by 2.91 m (6.3 cm/year) between 1963 and 2009 (Abate et al., 2015). The adjacent Megech River to the north has similarly experienced sedimentation of the old channel downstream of a bifurcation since 1998 (SMEC, 2008a). This means that sedimentation is observed in practically all rivers debouching in the Lake Tana from the Fogera plain.

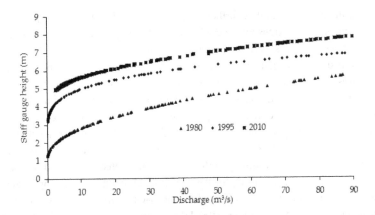

Figure 3.2. Ribb River staff level at the Lower gauging station with corresponding discharges (Data source: Ministry of Water, Irrigation and Energy, Ethiopia).

Confirmation of the sedimentation phenomenon was also given by the partial burial of several measuring stations in the region (Figure 3.3). The measurement of bed level around the old automatic water level recorder located near the Ribb Bridge shows that in 2016 the cylindrical station was buried by 2.4 m of sediment, which represents a rise of bed elevation of 4.2 cm per year. The study by Abate et al. (2015) indicated a rise of 1.4 m (2.6 cm/year) over 53 years (from 1959 to 2012) for the adjacent Gumara River. The calculated elevation change in floodplain may sustain uncertainties related to the natural settlement of the cylinder relative to the floodplain.

Figure 3.3. Automatic water level recorders installed in 1959 by the United States Bureau of Reclamation (USBR). The left and the middle are for the Ribb and Gumara Rivers located near the Ribb and Gumara Bridges, respectively. The right is for the Blue Nile River 4 km downstream of Chara-Chara weir (the outlet of the river from Lake Tana). This cylinder height was used as a reference to measure the developed levee, as it was not affected by sedimentation. $\Delta z$ is 140 cm, which is the height below the door (Abate et al., 2015).

The increase in river bed level is further supported by analyzing the computed reach-averaged annual sediment transport rates ($Q_s$) along the river. For the sake of simplicity, the representative river cross-section for each reach was assumed to be rectangular (see Table 3.2 for its dimensions), the flow was assumed as uniform, and the Chézy roughness coefficient was assumed constant along the river with a value of 35 $m^{1/2}$/s. Using the sediment transport Equations (3.3) and (3.4), applied to the period of available flow data, the average annual sediment transport rates were calculated as $6.96 \times 10^5$ $m^3$/year and $2.00 \times 10^5$ $m^3$/year for the Upper-I and Upper-II reaches, respectively, and $6.51 \times 10^4$ $m^3$/year and $5.70 \times 10^4$ $m^3$/year for the Upper Middle and Lower Middle sub-reaches, respectively. Assuming uniform flow and constant river dimensions to estimate the sediment transport rate for the Lower reach would give strongly uncertain results, as the river flow is highly affected by the backwaters caused by variations in Lake Tana and floodplain flooding (floodplains are not included in the simple sediment transport computations). The computation of sediment transport rate of the to the Ribb Bridge allows to conceptualize the longitudinal river bed profile changes, as depicted in Figure 3.4. The reduced sediment transport rate of the river in the downstream direction causes the river bed to rise as observed at the Lower gauging station (Figure 3.2).

Table 3.2. Reach-scale planimetric and geometric characteristics of Ribb River.

| Reach | Year | Length of Channel (km) | Valley Length (km) | Sinu osity (-) | Change of Sinuosity from 1957 (%) | Avg. River Width (m) | Avg. River Depth (m) | Valley Slope (%) | River Channel Slope (%) |
|-------|------|----|----|----|----|----|----|----|----|
| Upper-I | 1957 | 9.9 | 4.9 | 1.61 | - | - | | - | - |
| | 1980 | 10 | | 1.63 | 2.04 | - | | - | - |
| | 2006 | 10.4 | | 1.71 | 10.2 | - | | - | - |
| | 2012 | 9.6 | | 1.55 | −6.12 | - | | - | - |
| | 2016 | 9.7 | | 1.57 | −4.08 | 65 | 4.3 | 0.64 | 0.3 |
| Upper-II | 1957 | 20.8 | 13.3 | 1.71 | - | - | | - | - |
| | 1980 | 21.1 | | 1.74 | 2.26 | - | | - | - |
| | 2006 | 21.8 | | 1.79 | 7.52 | - | | - | - |
| | 2012 | 21.7 | | 1.78 | 6.77 | - | | - | - |
| | 2016 | 22.3 | | 1.83 | 11.28 | 58 | 4.8 | 0.21 | 0.12 |
| Middle | 1957 | 27.4 | 14.2 | 1.93 | - | - | | - | - |
| | 1980 | 25.2 | | 1.77 | −15.49 | - | | - | - |
| | 2006 | 24.7 | | 1.74 | −19.01 | - | | - | - |
| | 2012 | 24.8 | | 1.75 | −18.31 | - | | - | - |
| | 2016 | 25 | | 1.76 | −16.90 | 46 | 5.2 | 0.1 | 0.04 |
| Lower, (old) | 1957 | 22.8 | 15.3 | 1.49 | - | - | | - | - |
| | 1980 | 23 | | 1.5 | 1.31 | - | | - | - |
| | 2006 | 22.9 | | 1.5 | 0.65 | - | | - | - |
| Lower, (new) | 2012 | 20.1 | 13.4 | 1.5 | - | - | | - | - |
| | 2016 | 20.2 | | 1.51 | 1.73 | 38 | 5.5 | 0.05 | 0.037 |

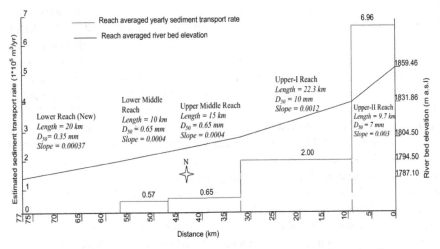

Figure 3.4. Sedimentation trend along the Ribb River based on estimated sediment transport rates. Distance of 0 km corresponds for dam site while 77 km is Lake Tana.

The average annual sediment transport rates in the old Lower river reach were calculated as $1.09 \times 10^5$ m$^3$/year and $1.04 \times 10^5$ m$^3$/year for the periods 1995–2001 and 2002–2008, respectively, if all discharges were conveyed by the river channel. The annual sediment transport rate result $1.09 \times 10^5$ m$^3$/year and $9.79 \times 10^4$ m$^3$/year for the periods 1995–2001 and 2002–2008, respectively, if the old Lower river channel was able to convey the discharge until bankfull. The two scenarios results in very similar sediment transport rates, which is due to the low frequency and the small duration of the flows that exceed the bankfull condition. These values are larger than the values obtained in the Upper and Lower Middle reaches. However, the actual values may be less than the estimated as it may be affected by the over bank flooding and back water effect of Lake Tana. These values were used to analyze the effect of Lake Tana level regulation on the river bed adjustment as described in Section 3.3.5 and the results are presented in Section 3.5.4.

### 3.4.3 Historical Channel Evolution

The river reach-scale characteristics in 1957, 1980, 2006, 2012 and 2016 are presented in Table 3.2, where negative values for sinuosity changes indicate length reduction. The river planform changes between 1957 and 2016 are shown in Figure 3.5 and mainly caused by bank retreat, cut-off formation, channel avulsion, and complete channel blockage.

The analysis of aerial photographs and satellite images shows that the study reach of the Ribb River reduced its length from 81 km to 77 km since 1957. In the Upper-I and Upper-

II reaches, the river has experienced only little planform changes, with −4.08% and 11.28% sinuosity changes resulting in a 0.2 km length reduction and 1.5 km length increment between 1957 and 2016, respectively. The rocky nature of the channel bed and banks in the Upper-I reach controls channel shape and width. The Middle reach reduced in length by 2.4 km, resulting in a sinuosity reduction of 16.9%. This was due to a channel avulsion event (Figure 3.5A) and a cut-off (Figure 3.5B), which occurred in the period 1980–2006. Another cut-off event (Figure 3.5D) occurred in the period 1957–1980. The small avulsion in Figure 3.5A reduced the channel length from 7.4 km to 6.3 km and the sinuosity from 1.41 to 1.20, while the cut-off in Figure 3.5D reduced the river channel length from 3.0 km to 1.4 km and the local sinuosity from 2.51 to 1.11. Generally, channel cut-offs and avulsions immediately increase the energy slope and the bed shear stress, and may coarsen the river bed materials (Frings et al., 2011). However, in the absence of further changes, the channel tends to restore its characteristics (slope, width, etc.) as before the shortening, with channel incision as a result. Embankments have been built in the Lower Middle sub-reach to reduce flooding. These embankments have also drastically reduced local channel migration. In the Lower reach, the channel experienced sedimentation, which resulted in complete river channel blockage starting 4 km downstream of the bridge in 2008 followed by channel of the new river channel is 2.6 km larger than that of the old river reach but has similar sinuosity (Table 3.2).

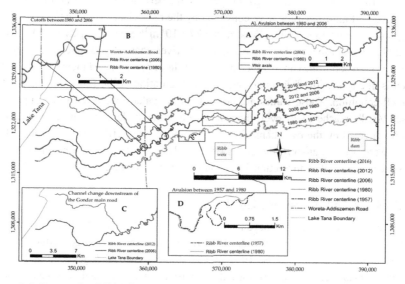

Figure 3.5. Successive Ribb River centrelines. (A) Avulsion developed between the years 1980 and 2006; (B) cut-off developed between the years 1980 and 2006; (C) complete channel change between the years 2006 and 2012; (D) cut-off developed between the years 1957 and 1980.

The temporal river planform changes occurred in the Middle reach in the period between 1980 and 2006 are shown using the aerial photographs 1957 and 1980 and the SPOT satellite images 2006 and 2016 (Figure 3.6).

The main channel width measured from the Google Earth satellite images of the year 2016 shows gradual narrowing in the downstream direction. This may be due to decreasing bank erodibility as the clay content of the bank material increases in downstream. In addition, the reduction of discharge in the downstream direction due to pump irrigation and flooding in the Lower reach may lead to channel narrowing as sediment may be deposited along the channel margins because of reduced transport capacity. It was not possible to derive the past channel width from historical satellite images because of poor resolution.

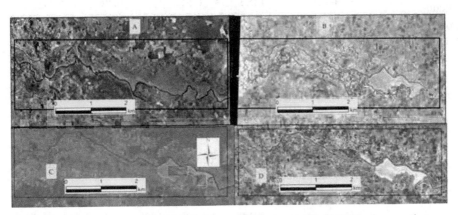

Figure 3.6. Aerial photographs and SPOT satellite images of the Middle river section downstream of the weir. (A, B) aerial photograph and river pattern of the year 1957 and 1980, respectively, and (C, D) SPOT satellite image and river pattern of the year 2006 and 2012, respectively.

## 3.4.4 River Slope Adjustment to Interventions

The long-term effects of sand mining and water extraction are assessed for the scenarios presented in Section 3.3.4 using the Equilibrium Theory. Table 3.3 lists the annual volume of river discharge for each scenario and the corresponding calculated reach-based annual sediment transport capacities, which are necessary to estimate the river slope ($i_0$) of the year 1980 using Equation (3.10).

Table 3.3. Annual average river discharge volume and calculated reach-averaged annual sediment transport capacity without and with interventions represented by Sc- 1 and Sc-2.

| Reach Name | Watershed Area (km²) | Average Annual River Discharge Volume ($\times 10^6$ m³) | | Reach-Averaged Annual Sediment Transport Volume ($\times 10^4$ m³) | | |
|---|---|---|---|---|---|---|
| | | Without Intervention | Scenario 1 | Without Intervention | Scenario 1 | Scenario 2 |
| Upper-I | 844 | 268.5 | 268.5 | 69.6 | 69.6 | 69.58 |
| Upper-II | 1162 | 370.0 | 370.0 | 20.0 | 20.0 | 20.02 |
| Upper Middle | 1381 | 440.4 | 425.3 | 6.51 | 6.49 | 5.84 |
| Lower Middle | 1592 | 478.5 | 458.6 | 5.7 | 5.66 | 5.09 |

The results listed in Table 3.4 show that water extraction for irrigation during the dry months has little or no effect on river bed slope. However, the river shows a response by decreasing its slope due to sand mining, considering that 6000 m³/year correspond to about 10% of the yearly reach-averaged sediment transport capacity.

Table 3.4. Calculated reach-averaged river bed slope in the year 1980 (before intervention) for the Middle River reach for Scenario 1 and Scenario 2.

| Reach | $i_\infty$, Slope in 2014 as Determined from 2014 ASTER DEM (%) | $i_o$, Estimated Slope of the Year 1980 for Scenario 1 (%) | $i_o$, Estimated Slope of the Year 1980 for Scenario 2 (%) |
|---|---|---|---|
| Middle (Upper and Lower) | 0.04 | 0.04 | 0.0426 |

## 3.4.5 Propagation of River Bed Level Adjustment to Lake Tana Regulation

Applying Equation (3.11) to analyze the Ribb River bed response to the Lake Tana level rise of 45 cm that was imposed in the period 1995–2001 (Figure 2.7) shows that, by 2008, a river bed rise of 22.5 cm may have propagated as far as 18 km upstream (upper limit) (i.e., after 13 years from the start of Lake Tana level regulation). The computed value of L is practically the same for the two channel conveyance scenarios, due to the very similar sediment transport rates (Section 3.4.2). Being the distance between the avulsion location and the river mouth 19 km (computed along the old channel), the results show that a rise of channel bed of approximately 20 cm may have reached the Ribb Bridge location at the time of channel avulsion. Instead, most probably the subsequent lowering of Lake Tana level that started in 2001 did not affect the location of the bridge. In 2008, retrogressive erosion resulting in bed level lowering of about 30 cm may have reached the distance of 9 km only.

At the location of the Ribb Bridge, in 2008, the bed level rise caused by Lake Tana regulation had thus the order of magnitude of 20 cm or less. This result indicates that the artificial rise of Lake Tana level may have been influenced the closure of the old Ribb River channel and the 2008 channel avulsion. However, the sedimentation caused by Lake Tana regulation appears negligible when compared to the observed total sediment deposition at the avulsion location of almost 3 m (Section 3.4.2).

## 3.5 DISCUSSION

### 3.5.1 River Channel Changes

The major planform change observed in the Ribb River system in the study period is the 2008 channel avulsion occurred 20 km upstream of Lake Tana. A slight contribution of Lake Tana level regulation on the rise of the river bed at this location is confirmed by the application of the methodology developed by De Vries (1975) (Section 3.4.5). The analysis shows that the 45 cm rise of lake level in 1995 could have caused an increase in bed elevation of about 20 cm at the location and time of the avulsion event. However, at the same location, the river bed level rose by about 3 m in the period 1980–2010 (Figure 3.2). This means that lake level regulation may have contributed to the closure of the old Ribb River channel which led to the avulsion event of 2008 but was not the main cause.

The largest part of the observed river bed aggradation has been caused by other factors that have impacted upstream sediment supply and downstream sediment transport capacity. The results of this study show that the sediment transport rate of the Ribb River decreases in the downstream direction (Section 3.4.2), which means that part of the sediment that is transported by the river in its upper parts is deposited in the Middle and Lower reaches. The downstream decrease in river sediment transport capacity is most probably aggravated by the backwater effects from Lake Tana that are not accounted for in the computations of the sediment transport rates.

In addition, LULC changes within the watershed may have increased the suspended sediment loads, which are not taken into account by sediment transport capacity formulas. Fine sediment settles in areas with low-flow velocities, such as floodplains and low-gradient reaches, such as the Lower reach of the Ribb River. Garede and Minale (2014) showed a 20% increase of cultivated land and a 10% and 21% decrease of bush land and forest cover of the Ribb watershed, respectively, between 1973 and 1995. Between 1995 and 2011 the extension of cultivated land remained approximately the same, whereas bush land and forest cover increased by just 1.5% and 6%, respectively. At the same time, the observed watershed precipitation and river discharge do not show any increasing or decreasing trends (Figure 2.8), this means that the erosive power of rain has not changed within the study time. Garede and Minale's findings, therefore, indicate that a possible increase of fine sediment supply may have started in the 1970s, due to LULC changes.

## 3.5.2 River Response to Sand Mining and Water Withdrawal

Sand mining activities are expected to have decreased the equilibrium longitudinal bed slope of the river since 1980. The slope may reduce further if the amount of sand mining increases, which is possible if the current increase of population and expansion of urbanization for the construction of buildings and other civil works was considered. If the current sediment mining increases to, for instance, 15% of the yearly sediment transports capacity of the Middle reach (which corresponds to extracting about 9000 m$^3$ of sand from the river bed every year); the river bed slope is expected to reduce to 0.036%. The operation of the planned Ribb Dam and diversion weir in the near future will reduce the sediment transport capacity in the downstream reaches and also the sediment inputs. The river will most probably respond by further decreasing its slope as a result of river bed erosion in the upper reaches, where increased bank height may lead to channel instability (Kondolf, 1997). Sediment mining combined with Ribb Dam operation may also create uneven river bed topography (Thornton et al., 2006; Frings, 2015) and coarsen the river bed material (Frings, 2015) which may result in an increased bed roughness and reduced flow velocity. On the long term, deficiency of fine sediment supply to the downstream reaches, and sediment mining may lead to shoreline retrogression (Jiang et al., 2013).

The analysis shows that water extraction during the dry months does not affect the river bed slope (Table 3.4). This may be due to the small river discharge (on average 1.9 m$^3$/s); so that the direct impact on the river morphology is minimal. However, the current reduction of flow discharge due to water withdrawal creates conditions that are favorable to sand miners who can easily excavate more sediment from the bar tops. This further contributes to channel slope decrease. In the future, the very small environmental flows (0.15 to 0.17 m$^3$/s) will have no negative effects on sediment mining, which is thus expected to continue or even increase.

## 3.5.3 Effects of the Embankments

Continued embankment construction in the Lower river reach will contain the water flow and reduce the already limited lateral channel migration (Surian, 1999; Frings, 2015). The volume of water through the river channel will increase whereas floodplain water and sediment flow will decrease. The latter might have a negative impact on local agriculture. Channel embankment will result in higher flow velocity and sediment transport capacity (Jansen et al., 1979; Frings, 2015), which will reduce the observed on-going sedimentation phenomenon. Moreover, the embankments construction at the upstream reach will increase the propagation speed and cause severe flooding damage at the downstream reaches (Popescu et al., 2010). The villagers living at the lower reach of the study area where embankments are not constructed (especially along the Ribb River) claim the existence of high amount of flooding than before. The operation of Ribb Dam will further reduce sedimentation in the future. Finally, increased bank height and flow confinement may increase the rate of bank erosion and the risk of bank failure (Li et al., 2007).

### 3.5.4 Application of the Equilibrium Theory

Using the Equilibrium Theory, including the formula of de Vries that is based on the same approach leads to highly uncertain results, and, for this reason, all quantifications should be only considered for comparison between scenarios to identify increasing or decreasing trends. The theory was developed for sand-bed rivers, for which the threshold velocity for sediment particle motion is close to zero. This is not true for gravel-bed rivers, where the existence of a clear threshold results in limitations of river bed degradation due to armoring. Equilibrium Theory was applied to the Middle reach of the Ribb River to infer the longitudinal channel slope prior to human interventions (reference year: 1980 AD). The bed material of the Middle reach is made by sand (70%) and gravel (30%), the median grain size being 0.65 mm. In this area, bed armoring can indeed limit river bed degradation. The results show that the river most probably experienced a temporal decrease in bed slope due to sand mining, but no changes occurred due to water extraction. Considering the strong limitations of the approach used outside its applicability range, the computed slopes for 1980 AD should be considered only to infer past trends (increase or decrease).

The formula by de Vries was used to roughly assess whether retrogressive sedimentation due to Lake Tana level rise had reached the location of channel avulsion by 2008. The Lower reach has a sand bed and the formula is therefore applicable. Nevertheless, the results are only indicative, due to the strongly simplified approach.

### 3.6 CONCLUSIONS

The analysis of the morphodynamic trends of the Ribb River between 1980 and 2016 presents the response of a medium-size low-land river belonging to the Lake Tana sub-basin to anthropogenic drivers that are typical of the region: sand mining, water extraction, embankment construction, and downstream boundary alteration. The study examines the responses of the river to these interventions by image analysis and by the application of physics-based analytical methods.

In particular, having subdivided the last 77 km of the river in four reaches, the work provides the current reach-scale sediment budgets. The results show a reduced sediment transport capacity in the downstream direction, a condition that leads to progressive sediment deposition. This is confirmed by the river bed level rise of almost 3 m observed near the Ribb Bridge, about 20 km upstream of Lake Tana (river mouth), in the 30 year period (1980–2010). The complete silt up of the river channel 4 km more downstream, which resulted in river avulsion in 2008, was most probably the result of the observed sediment transport capacity decrease, aggravated by the backwater effects related to the artificial rise of Lake Tana level for hydropower production, which have further reduced the sediment transport capacity of the flow in the downstream direction. The analysis of the upstream propagation time of river bed rise from Lake Tana confirms the hypothesis that lake regulation contributed to channel avulsion.

The reduction of sediment transport capacity in the downstream direction appears to be mainly caused by the upstream sediment mining and water withdrawal for irrigation, the former by reducing the longitudinal river bed slope of the river and the latter by directly reducing the water flow and its transport capacity. Sedimentation and downstream decrease of the water discharge are also responsible for the observed width reduction in the downstream direction.

By providing a full quantitative description of the river prior to the construction of the Ribb Dam, 77 km upstream of the river mouth, and related diversion weir, 30 km more downstream, the results of this work will allow for estimating the effects of the future operations on river morphology and dynamics.

# 4

# LONG-TERM EFFECTS OF RIBB DAM OPERATION FOR IRRIGATION ON DOWNSTREAM RIVER REACHES[2]

This chapter assesses the applicability of an analytical method for quick assessments of the long-term morphological effects of different dam operations. The idea is to apply the method in feasibility studies to quickly identify the least morphologic-impacting operation scenario. The outcome of the analytical method is compared to the results of a calibrated one-dimensional morphodynamic model.

---

[2] This chapter is based on:
Mulatu, C. A., Crosato, A., Langendoen, E. J., Moges, M. M., & McClain, M. (2020). Long-term effects of dam operations for water supply to irrigation on downstream river reaches. The case of the Ribb River, Ethiopia. *The International Journal of River Basin Management*. doi:10.1080/15715124.2020.1750421.

## 4.1 INTRODUCTION

Dams are vital for the economic development of regions due to the need of storing water for different activities, such as flood control, power generation and irrigation. However, dams have major impacts on downstream river reaches. They change the discharge regime; in particular, the timing, duration, and magnitude of low and high flows (Williams and Wolman, 1984; Surian et al., 2009; Grant, 2012; Lobera et al., 2015; Rubin et al., 2015). In addition, dams also change the sediment regime, since they trap almost all the incoming bed load and a portion of the suspended load; the reservoir's trapping efficiency depends, among others, on the size and shape of the reservoir and its operation (Brune, 1953; Basson, 2004; Teruggi and Rinaldi, 2009; Kondolf et al., 2014a; Kondolf et al., 2014b). In response to the changes in discharge and sediment regimes, the channel downstream of the dam adjusts its morphology by altering its planform, slope, width, depth, and sediment characteristics through time.

Both channel bed degradation and aggradation are the primary responses that have been observed downstream of dams (Williams and Wolman, 1984; Schmidt and Wilcock, 2008; Lobera et al., 2015). For example, up to 7.5 m of river bed lowering has been observed downstream of the Hoover Dam on the Colorado River, United States of America; degradation extended to 21 km downstream 6 months after dam closure, 28 km after 1 year, 50 km after 2 years and over 120 km after 5 years (Williams and Wolman, 1984). Degradation may eventually change the river planform, from braided to meandering (Kondolf, 1997). River bed material coarsening and armor layer formation may continue until the released flow is unable to entrain the bed material (Williams and Wolman, 1984; Holly Jr and Karim, 1986; Kondolf, 1997). On the other hand, if the regulated flows are not able to move sediments supplied by tributaries, the river bed may aggrade (Williams and Wolman, 1984). Long term operation of dams may also change the concavity of the downstream river bed profile and result in bed material fining (Nones et al., 2019; Varrani et al., 2019). Channel widening, bar formation, braiding, increased sinuosity (Shields Jr et al., 2000) and reduced channel conveyance capacity (Williams and Wolman, 1984; Sanyal, 2017) are further frequent consequences. Generally, the overall morphological impacts of dam construction depend on the river size and reservoir capacity, pre- and post-dam hydrologic regimes, environmental setting, initial channel morphology and dam operation (Williams and Wolman, 1984; Church, 1995; Petts and Gurnell, 2005; Graf, 2006; Schmidt and Wilcock, 2008; Yang et al., 2011; Li et al., 2018).

A generalized prediction of dam-induced impacts on downstream channel morphology is difficult, as the drivers and processes are site-specific and complex interactions exist among these drivers (Graf, 2006). A number of assessment methods have been developed with varying complexity. Williams and Wolman (1984) performed an empirical analysis using data of different river systems. Petts (1979), Brandt (2000) and Petts and Gurnell (2005) developed conceptual models. Grant et al. (2003), Curtis et al. (2010) and Schmidt and Wilcock (2008) developed channel response predictive models for general applications. In addition, numerous one-dimensional (1D), two-dimensional (2D) and three-dimensional (3D) morphodynamic codes are available to analyze dam-induced morphologic changes (Khan et al., 2014; Omer et al., 2015). 1D models have advantages for large-scale studies, as they are less data intensive and computationally efficient while providing an acceptable general representation of the river system (Berends et al., 2015). These models are appropriate to study the dam-induced effects at the preliminary and feasibility stages of dam design and to optimize dam operation scenarios (Van der Zwet, 2012; Nones et al., 2014). They also allow to analyze the time evolution of river morphology, and can establish the duration to reach a new equilibrium, as well as the temporal evolution of local river characteristics like sediment transport rates, water level and river bed elevation changes.

The long-term morphological effects of different dam operations on downstream river reaches can also be assessed using simple, easy to apply, analytical methods, for example the Equilibrium Theory (ET) developed by Jansen et al. (1979). Such analytical methods have the following advantages compared to 1D (and greater) computer models: much fewer data are required; reduced setup, calibrate, and run time; and therefore reduced costs. The ET method estimates the theoretical river bed slope, which would be attained a long time after impoundment, caused by the changed discharge and sediment transport regimes. This method, however, is unable to predict the time required to reach the final equilibrium state and the transition phases of the morphological development.

In the absence of river reaches with historical pre- and post-dam bed profile data, it is impossible to assess how well the above-mentioned methods assess the morphological changes. Additionally, if historical data related to the pre-dam period are lacking in systems that have been impounded in the past, it is unknown if the downstream reaches have attained a condition close to a new equilibrium. These limitations are especially true for rivers in Africa, like those of the Upper Blue Nile Basin, Ethiopia, where most dams have been constructed during last decades (Koga dam), or are planned (Gilgel Abay and Gumara dams) or under construction (Ribb and Megech dams). Determining the least morphologically impactful dam operation scenario is the critical to the regional use of water resources.

The overall goal of the presented study is to establish whether the ET method allows identifying the least-impactful dam operation scenario so it can be used in the preliminary dam design phases. The Ribb River reach downstream of the dam is selected for the case-study. The bed of the study reach is dominated by the gravel and sand formations for which hydrologic and geomorphic data are available (Mulatu et al., 2017; Mulatu et al., 2018). To assess the applicability of the ET method, its results are compared to those from a calibrated 1D morphodynamic model developed using the SOBEK-RE software. The software has been applied for similar purposes by Van der Zwet (2012) for the White

Volta River, Ghana, to analyze the morphological change due to damming. He found river bed degradation just downstream of the dam propagated in a downstream direction through the time of dam operation for the alluvial deposit river reach.

## 4.2 STUDY CASE: THE RIBB RIVER

The river system downstream of the Ribb Dam was divided into four reaches as described in Section 2.6.1 (the Upper-I, the Upper-II, the Middle and the Lower) based on the presence of natural and man-made river bed fixations (rock outcrops, diversion weir, and bridge) (Figure 2.9). The reaches differ according to their bed material, average river bed slope, and channel width, as well as anthropogenic interventions (embankment construction, water abstraction and, sand mining). Planimetric, geometric and granulometric characteristics of the study reaches were analyzed by Mulatu et al. (2018) (Chapter 3). The main parameters needed for this study are summarized in Table 4.1.

Table 4.1. Reach-average characteristics of the Ribb River below the Ribb Dam.

|  | Reach length (km) | Reach-averaged bankfull width (m) | Reach-averaged bankfull depth (m) | Reach-averaged channel bed slope (%) | Median grain size $(D_{50})$ (mm) | Mean grain size $D_m$ (mm) | $D_{90}$ (mm) |
|---|---|---|---|---|---|---|---|
| Upper-I | 10 | 65 | 4.3 | 0.30 | 7.0 | 9.7 | 12 |
| Upper-II | 22 | 58 | 4.8 | 0.12 | 7.0 | 7.5 | 11 |
| Middle | 25 | 46 | 5.2 | 0.04 | 0.65 | 3.0 | 5.5 |
| Lower | 20 | 38 | 5.5 | 0.037 | 0.35 | 0.5 | 0.6 |

## 4.3 MATERIAL AND METHODS

A physics-based analytical method derived from the ET, developed by Jansen et al. (1979), and a 1D model constructed using the SOBEK-RE software (http://sobek-re.deltares.nl/index.php) was applied to estimate the expected long-term morphological changes of the Ribb River downstream of the Ribb Dam. Several dam operation scenarios were considered. The results of the two applications were then compared to assess the applicability of the ET for the preliminary dam designs. The Ribb River has distinct sand-bed (Lower and Middle) and gravel-bed (Upper-I and Upper-II) reaches; the latter may present a limit to bed degradation due to bed armoring. The detailed description of the methods, the required input data and the boundary conditions for the ET and the 1D model are presented in the next sections.

## 4.3.1 Data Sources

The required data for the study were collected from the literature and provided by Ethiopian agencies. For instance, the current reach-averaged river bed slopes, cross-sections, and the reach-averaged river bed-materials characteristics were derived from Mulatu et al. (2018), who collected the required data and bed-material samples to analyze the current morphodynamic trends of the Ribb River, i.e. before dam construction. The daily time series discharge at the Lower and Upper gauging stations and the water levels at the Lower gauging station covered the period 1995-2010 were collected by the MoWIE. The discharge data were used to compute the reach-averaged pre-dam sediment transport capacity and for water balance analysis, while the water levels was used for the calibration and validation of the 1D model. The monthly average rainfall and evaporation at the reservoir surface of the same period were obtained from the Ribb Dam feasibility study and design documents (Figure 4.1). These documents, provided by the MoWIE, were also used as a data source to develop realistic dam operation scenarios.

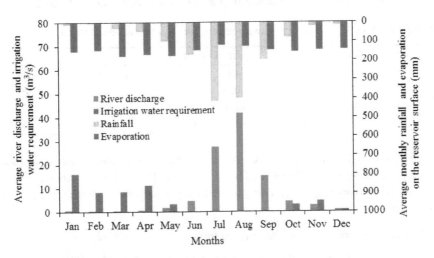

Figure 4.1. Comparison of average monthly river discharge at the Upper gauging station (m³/s), irrigation water requirement (m³/s), monthly average rainfall over the reservoir (mm), and monthly average estimated evaporation at the reservoir surface (mm). Data source: Study document for the feasibility of Ribb Dam (WWDSE and TAHAL, 2007).

## 4.3.2 Dam Operation Scenarios

The reservoir operation was designed to irrigate 15,000 ha of command area for the specified monthly irrigation water requirement and environmental releases (WWDSE and

TAHAL, 2007), Figure 4.1. However, the dam may be operated in the future for a larger or smaller command area, deviating from the design. For example, the Ribb Dam Feasibility Study document (WWDSE and TAHAL, 2007) suggests that the stored water may be released to irrigate 19,925 ha of command area, which is 32.8% greater than the design.

Taking this into account, three scenarios were considered: Scenario-I (Sc-I) corresponding to a command area of 13,500 ha (10% smaller from the design); Scenario-II (Sc-II) corresponding to a command area of 15,000 ha (equal to the design); and Scenario-III (Sc-III) corresponding to a command area of 19,925 ha (32.8% greater from the design). Dam releases to irrigate the proposed command areas (irrigation water requirements) for each scenario were provided as monthly discharges (Table 4.2). However, the actual water release may not correspond to either of the studied scenarios, as it will depend on the actual cropping pattern and the amount of rainfall in the area. In all dam operation scenarios, the dry month discharge downstream of the weir will constitute the ecological flow, which is 0.17 $m^3$/s or lower (BRLi and MCE, 2010).

Table 4.2. Monthly crop water requirements for the dam operation scenarios.

| Scenario | Monthly crop water requirements ($m^3$/s) | | | | | | | | |
|---|---|---|---|---|---|---|---|---|---|
| | Jan | Feb | Mar | Apr | May | Jun to Sep | Oct | Nov | Dec |
| Sc-I | 14.62 | 7.7 | 7.81 | 10.26 | 3.02 | 0 | 2.67 | 4.06 | 0.86 |
| Sc-II | 16.24 | 8.56 | 8.68 | 11.4 | 3.36 | 0 | 2.97 | 4.51 | 0.95 |
| Sc-III | 21.57 | 11.37 | 11.53 | 15.14 | 4.46 | 0 | 3.94 | 6.00 | 1.26 |

Once the reservoir is full, the wet season excess volume of water above the Normal Pool Level (NPL) elevation will be released by the spillway. The spill-over discharge is here determined by applying a reservoir water balance based on the continuity equation (Equation 4. 1).

Reservoir volume change = Average inflow − Average outflow          (4.1)

The river discharge and rainfall over the reservoir are considered as inflow, whereas seepage, evaporation, environmental flow, and irrigation water requirements are considered as outflow. WWDSE and TAHAL (2007) estimated the seepage loss to be 25% of the monthly environmental flow.

The volume of rainfall and evaporation over the reservoir are surface-area dependent and hence, an Elevation - Area - Volume relationship of the reservoir was developed from the available topographic data of the reservoir area. For simplicity, a monthly time step was considered, even though this will reduce the outflow peaks and affect the results of yearly sediment transport capacity. The reservoir volume loss due to sediment accumulation through time was not considered for the water balance analysis. WWDSE and TAHAL (2007) estimated a reservoir storage loss of 0.22% per year. As sediment supply to the reservoir is dynamic and mainly depends on LULC, population density, climate change, etc., there exists significant uncertainty in the expected storage loss and therefore its possible impact on the water balance analysis. Though note, reservoir sediment accumulation will reduce the reservoir volume over time and increase the spill-over discharge events, which will result in an increased downstream morphological activity.

The pre-dam discharges were derived using the time-series discharge data at the Upper and the Lower gauging stations. The irrigation water requirements, the environmental releases, the lateral inflow, and the spill-over discharge were used to derive the post-dam downstream discharge characteristics.

### 4.3.3 Sediment Transport Capacity

As described by Mulatu et al. (2018), the Upper-I and the Upper-II reaches of the river system are dominated by gravel, while the Middle and the Lower reaches by sand. The Meyer-Peter and Müller (1948) sediment transport formula (Equation 3.3) revised by Wong and Parker (2006), assuming no bedforms ($\mu=1$), and the Engelund and Hansen (1967) sediment transport formula (Equation 3.4) were selected to compute the sediment transport capacity of the gravel and sand dominated river reaches, respectively. These equations were developed for gravel and sand dominated systems, respectively, and selected based on data availability.

The Lower reach can have significant overbank flow during the rainy season (SMEC, 2008a; Mulatu et al., 2018). Therefore, the measured discharge at the Lower gauging station is not representative for the entire reach. Hence, the computation of the reach-averaged sediment transport capacity for this reach may have some uncertainty. For this reason, the Lower reach is excluded from the analytical model application.

The pre-dam reach-averaged sediment transport capacity of the river was estimated by Mulatu et al. (2018) using daily river discharges. Here, however, as the post-dam outflow discharge was given on a monthly basis, the pre-dam sediment transport using a monthly average discharge was computed. In this way the annual sediment transport rate of the Ribb River might be underestimated, but this is still acceptable considering that the results are used to compare scenarios, all based on monthly discharges.

### 4.3.4 Reach-Scale Morphodynamic Equilibrium Theory

A river reach can be considered in morphodynamic equilibrium if the on-going reach-scale morphological changes can be considered negligible. This is true only if the volume of sediment entering the river reach equals the volume of sediment leaving the same river reach during a chosen time interval (Jansen et al., 1979; Bolla Pittaluga et al., 2014). Any new interventions on a river reach in morphodynamic equilibrium trigger a new trend of erosion and deposition, which ultimately leads to a new reach-scale equilibrium characterized by different values of the morphological variables, such as reach-averaged longitudinal slope, channel width, sediment size and bed roughness (Gurnell et al., 1994; Legleiter, 2014). This means that the river adapts its longitudinal bed slope and the other reach-averaged characteristics to the new external forcing, and this adaptation occurs during a transition period which can be rather long (tens to hundreds of years), depending on sediment transport rate and size of the river channel (Knighton, 1998; Wu et al., 2012).

The Equilibrium Theory was developed by Jansen et al. (1979). The theory compares two reach-scale morphodynamic equilibrium conditions: one before (old equilibrium condition) and the other one a long time after a change of forcing has occurred (new equilibrium condition). The old and new equilibrium states are characterized by their longitudinal bed slope, channel width, water discharge and sediment transport regimes. The method was used here to estimate the theoretical river bed slope attained a long time after impoundment due to the changed discharge and sediment transport regimes. The theory has been recently extended by, for instance, (i) Li et al. (2014) to predict the equilibrium bed profiles and water discharge for water diversions, (ii) Blom et al. (2017) to determine the equilibrium channel geometry of the alluvial river for variable flow relating with the channel width, bed material texture and channel slope, (iii) Bolla Pittaluga et al. (2014) to investigate the reach-scale quasi-equilibrium bed profile by applying a 1D morphodynamic model, and (iv) Lanzoni et al. (2015) to develop a 1D morphodynamic model to estimate the formative discharge that caused the observed (current) river bed slope of the Po River, Italy.

The methodology is based on the following major assumptions: (i) the river is fully alluvial, (ii) the river flow is uniform with a given width and a large width to depth ratio, (iii) the Chézy coefficient, the sediment composition and the degree of non-linearity of the sediment transport formula remain unchanged, and (iv) the river morphological response is based on the derivation of longitudinal slope and representative water depth.

The methodology was developed by combining the continuity equation of water (Equation 3.5), the momentum equation for water, reduced to Chézy's equation for steady uniform flow and for large flow width-to-depth ratios (Equation 3.6), the simplified sediment transport capacity formula expressed as a power law of flow velocity (Equation 3.7) and the sediment balance equation (Equation 3.8) as presented in Section 3.3.4.

## Equilibrium bed slope for gravel-dominated river reaches

For the gravel-bed reaches, the average yearly sediment transport capacity of the river is computed using Equation 3.3, based on monthly discharges. Sediment is transported by the water only if the flow velocity exceeds the critical value for particle motion.

According to the characterization of dams by the International Commission on Large Dams (ICOLD, 1998), the Ribb Dam will store all of the incoming bed loads and the majority of suspended loads. This means that the future sediment input to the downstream reaches can be considered negligible. Post-dam released discharge values, for which the corresponding flow velocity exceeds the limiting value for sediment entrainment ($u_c$), will have the capacity to entrain sediment from the river bed and transport it downstream. Upstream bed erosion results in decreasing bed slope. The process continues until the final equilibrium bed slope that yields zero sediment transport is attained. At this point, the river bed material becomes immobile for all discharges, and in particular for the largest dam released discharge. Hence, the maximum dam released monthly discharge was used here to estimate the final river bed slope which will be attained after long years of dam operation. This is the slope for which the sediment transported by the highest discharge has become zero (achievement of critical conditions). The post-dam water depth at the largest released discharge ($h_{M\infty}$) and zero sediment transport can be obtained from Equation 3.3 combined with Equation 3.5 and Equation 3.6 assuming the bed roughness, channel width and sediment size remain unchanged.

$$h_{M\infty} = \frac{Q_{wM}}{BC\sqrt{(0.047\Delta D_m)}}$$
(4.2)

in which $Q_{wM}$ is the highest monthly discharge released by the dam (m$^3$/s).

The final river bed slope ($i_\infty$) which yields zero sediment transport is obtained from Equation 3.3, and is written as:

$$i_\infty = \frac{0.047\Delta D_m}{h_{M\infty}}$$
(4.3)

Assuming constant Chézy coefficient and grain size is a limitation of the approach as they might both increase after the start of dam operation (Williams and Wolman, 1984; Grant, 2012). However, as described by Di Silvio and Nones (2014), morphodynamic results obtained by using uniform bed materials are sufficiently significant for the analysis of future morphodynamic response.

## Equilibrium bed slope for sand dominated river reaches

The total yearly sediment transport capacity for the sand-bed Middle river reach was derived using the Engelund and Hansen (1967) sediment transport formula, Equation 3.4. Combining the continuity (Equation 3.5) and momentum (Equation 3.6) equations for water and the sediment transport (Equation 3.7) and sediment balance (Equation 3.8) equations, the slope before and after dam operation for variable discharge (here monthly average) are given in Equation 4.4 and Equation 4.5, respectively. These equations are derived considering the monthly discharge of the river with the same probability density of $\frac{1}{12}$, in which k represents the specific month.

$$i_0 = \frac{Q_{s0}^{3/b_0} B^{(1-3/b_0)}}{a_0^{3/b_0} C_0^2} \left( \frac{1}{\sum_{i=1}^{12} \frac{1}{12} Q_{w0k}^{b_0/3}} \right)^{3/b_0}$$ (4.4)

$$i_\infty = \frac{Q_{s\infty}^{3/b_\infty} B^{(1-3/b_\infty)}}{a_\infty^{3/b_\infty} C_\infty^2} \left( \frac{1}{\sum_{i=1}^{12} \frac{1}{12} Q_{w\infty k}^{b_\infty/3}} \right)^{3/b_\infty}$$ (4.5)

In Equation 4.4 and Equation 4.5, the subscripts "0" and "∞" indicate the values at the initial (just before dam operation) and final (a long time after dam operation) equilibrium states, respectively.

The ratio between river bed slope at the final and at the initial equilibrium states for variable discharge, assuming constant sediment transport exponent, b, sediment transport proportionality coefficient, a, river width, B, and Chézy coefficient, C, is given by Equation 4.6. This equation was used to compute the final equilibrium state of the river bed slope for the Middle river reach.

$$\frac{i_\infty}{i_0} = \left( \frac{Q_{s\infty}}{Q_{s0}} \right)^{3/b} \left( \frac{\sum_{i=1}^{12} Q_{w0k}^{b/3}}{\sum_{i=1}^{12} Q_{w\infty k}^{b/3}} \right)^{3/b}$$ (4.6)

**Application of Equilibrium Theory to the Ribb River**

The application of the analytical method allows to assess the theoretical long-term response of the Ribb River due to altered discharge and sediment transport regimes caused by prolonged dam operation. The river reaches are assumed in a state of morphodynamic equilibrium at the moment of the intervention (i.e. no sediment accumulation or losses). This might not be true for the Ribb River, as there is a reduction in sediment transport capacity in the downstream direction along the reaches, leading to progressive aggradation and channel blockage starting 4 km downstream of the Ribb Bridge (Mulatu et al., 2018).

## 4.3.5 Morphodynamic Model Application

**Model description**

SOBEK-RE is a 1D modelling tool for open channel networks computing water flow, sediment transport (uniform and graded sediment) and morphological changes for compound river channels, considering main channel and floodplains separately (http://sobek-re.deltares.nl/index.php/documentation). The model comprises three modules: (i) the water flow module, described by the continuity (Equation 4.7) and momentum (Equation 4.8) equations for water; (ii) the sediment transport module, which allows selecting among five standard sediment transport capacity formulas plus a user-defined one; and (iii) the morphological module, describing the bed level adaptation through time (Equation 4.9) based on the sediment balance equation (Exner, 1925). To prevent any differences between the ET method and 1D model to be caused by the selected sediment transport equations, the sediment transport formulae of Meyer-Peter and Müller (1948), revised by Wong and Parker (2006), and Engelund and Hansen (1967) were also used in the SOBEK-RE to compute the sediment transport capacity for the gravel and sand dominated reaches, respectively.

$$\frac{\partial A}{\partial t} + \frac{\partial Q_w}{\partial x} = 0 \tag{4.7}$$

$$\frac{\partial Q_w}{\partial t} + \frac{\partial}{\partial x}\left(\alpha \frac{Q_w^2}{A}\right) + gA\frac{\partial h}{\partial x} + \frac{gQ_w|Q_w|}{C^2 RA} = 0 \tag{4.8}$$

$$(1-\varepsilon)\frac{\partial Z_b}{\partial t} + \frac{\partial q_s}{\partial x} = 0 \tag{4.9}$$

in which A is the wet cross-sectional surface area (m$^2$), t is time (s), x is the longitudinal distance (m), $\alpha$ is the Boussinesq coefficient (-), R is the hydraulic radius (m), $\varepsilon$ is the bed material porosity (40% for uniform sand), and $Z_b$ is the bed level (m). The hydrodynamic model simulation is based on implicit time discretization, in which the stability of the solution does not depend on the Courant number, though this will influence the accuracy of the solution (Ali, 2014). On the contrary, the time step for the solution of the sediment continuity equation must satisfy the Courant number ($\sigma$).

$$\sigma = \alpha_c c \frac{\Delta_t}{\Delta_x} \leq 1 \tag{4.10}$$

where c is the bed disturbance celerity (m/s), $\Delta_t$ is the morphological time step (s), $\Delta_x$ is the grid size (m) and $\alpha_c$ is the stability factor to prevent non-linear instability (taken as 1.01, based on SOBEK-RE user manual).

The Lower reach has several bifurcations close to Lake Tana, but SOBEK-RE model simulations was done assume a single, 77 km long river (four reaches) from the dam site to Lake Tana.

## Ribb River model construction

Model construction includes model setup, calibration and validation. For this study, the model was constructed and executed for three different dam operation scenarios (Section 4.3.2). Reach-averaged river bed slope, channel dimensions (width and depth, assuming a rectangular channel), mean grain size, Chézy coefficient, and upstream and downstream boundary conditions were the required input parameters as described in Section 4.3.1 and listed in Table 4.1. The model development was based on the assumption of uniform bed material for each study reach. The general schematization of the SOBEK-RE model for the Ribb River is shown in

Figure 4.2. All model simulations cover 1,500 years, as this period is assumed to be long enough to obtain a new morphodynamic equilibrium of the river reaches.

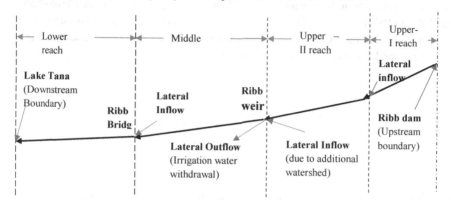

Figure 4.2. 1D model schematization with boundary conditions and location of hydraulic structures. Lateral inflow is the discharge due to additional watershed area, while the lateral outflow is the water withdrawal to the irrigation command areas by the diversion weir.

The measured average monthly river discharge and the average monthly dam releases (sum of irrigation demand, environmental flow, and spill-over discharge) were used as model input at the upper boundary to simulate the pre- and post-dam cases, respectively. A constant (time-averaged) Lake Tana water level was imposed at the downstream boundary for all simulation scenarios, even though this is variable and creates alternating local river bed erosion and deposition during the low and high stages, respectively. The long-term character of the study justifies using a constant level. The sediment transport rates of the Upper-I reach computed using Wong and Parker (2006)'s formula was used as the upstream boundary condition for the pre-dam case. For the post-dam cases, the sediment load at the upper boundary was set to zero, assuming that the dam blocks bed material movement to the downstream reaches. Discharge increment at the start of the Upper-II, Middle and Lower reaches due to the additional watershed area, and the water withdrawal for irrigation at the start of the Middle reach were introduced as lateral flows. Due to lack of data, the model neglects lateral sediment inputs by ephemeral tributaries, which may enter the river system during the rainy season. However, model sensitive to sediment supply was assessed by considering the release of 20% of the sediment transport capacity of the reaches to the downstream river from the dam as lateral sediment input (the released sediment volume depends on the dam operation scenarios). All model simulations cover 1,500 years, as this period is assumed to be long enough to obtain a new morphodynamic equilibrium of the river reaches, considering that gradually decreasing river slope reduces the sediment transport capacity, which might considerably slow the morphological changes.

The rock outcrops located at the end of Upper-I reach (around 1.2 km long) were introduced in the model as a non-erodible layer. Hence, in this river section no bed erosion occurs, but there may be some sediment deposition. The existence of this fixed river section controls the pre- and post-dam slope of the reaches. This section of the river cannot be taken into account by the ET, which is based on the assumption of an alluvial river without any geological constraints. In addition, assuming uniform bed material over the depth may also yield uncertain results, as there may be rock formation beneath the alluvial layers.

The diversion weir, located at the upstream boundary of the Middle reach, was included in the model as a structure. The weir temporarily affects the sediment movement causing aggradation and degradation at the upstream and downstream reaches (Jansen et al., 1979; Lanzoni et al., 2015; Ahn et al., 2017), elongating the time that is necessary for achieving equilibrium (Jansen et al., 1979; Lanzoni et al., 2015).

The bed roughness, in terms of Chézy's coefficient, was selected as a calibration parameter to compute the water depths and compare with the measured values of the months from May to October 2007 at the Lower gauging station. Based on visual characteristics of the study reach, Chézy coefficient values ranging from 30 $m^{1/2}$/s to 40 $m^{1/2}$/s were considered. The runs were carried out assuming that the upstream effects of backwater and floodplain flow are negligible, which may occur in the Lower reach if the discharge is above bank full discharge (110 $m^3$/s), Section 3.4.1. The Root Mean Square Error (RMSE), Equation 4.11, which measures the difference between the observed values and the model outputs, was used to identify the best value of Chézy coefficient. RMSE value close to zero indicates a good fit.

$$RMSE = \sqrt{\frac{\sum_{i=1}^{n}(X_i - X_0)^2}{n}}$$

(4.11)

where $X_i$ and $X_0$ are the measured and the model output (estimated) water levels and n is the number of samples. For validation, the calibrated model was run from May to October 2008. The computed water levels at the Lower gauging station were then compared to those measured, and RMSE was calculated.

The accuracy of the computed sediment transport capacity using a sediment transport formula may be limited due to the many uncertainties related to the physical characterization of the river reach and its hydraulic regime (Schmidt and Wilcock, 2008). Sensitivity analyses should thus be carried out to study the effects of varying the value of specific variables characterized by strong uncertainty (Hamby, 1994). In this study, a sensitivity analysis was carried out to analyze the effects of grain size on bed load in the Upper-I reach and on river bed degradation near the toe of the dam after 100 years for the range of $D_m$ values from 6 to 8 mm. The obtained results were then compared with the results of the base-case scenario (Sc-II).

## 4.4 RESULTS

### 4.4.1 Pre and Post-Dam River Discharges

The average monthly excess discharge that is expected to spill-over from the reservoir, computed from the reservoir water balance (Equation 4.1), is shown in Table 4.3. On average, the reservoir is expected to be full at the end of July and the spill-over to continue until October. In all dam operation scenarios, the total yearly volume of water flow in the Upper-I and Upper-II reaches will not be affected by the dam, except for the first year to fill the reservoir.

Table 4.3. Average monthly spill-over discharge for the three dam operation scenarios.

| Scenarios | Average monthly spill-over discharge (m³/s) | | | |
|---|---|---|---|---|
| | November to July | August | September | October |
| Sc-I | 0 | 32.4 | 15.4 | 1.4 |
| Sc-II | 0 | 27.2 | 15.4 | 1.1 |
| Sc-III | 0 | 9.8 | 15.4 | 0.1 |

The average pre- and post-dam monthly discharge hydrographs for the downstream river reaches are shown in Figure 4.3. The pre-dam hydrograph is directly related to the amount of rainfall and the watershed properties and shows one peak discharge period per year. After regulation, the Upper-I (Figure 4.3A) and Upper-II (Figure 4.3B) reaches will experience two peaks per year. The wet period peak is due to excess discharge from the reservoir, while the dry period peak is due to the maximum water release from the reservoir for irrigation. For the Sc-III dam operation the wet season peak in the Upper-I reach is shifted by one month (from August to September) and reverted back in the Upper-II reach. This is due to the additional discharge from the ungauged watershed. On average, the Sc-II dam operation reduces the unregulated wet period peak discharge by 34%, 25%, 21% and 21% for the Upper-I, Upper-II, Middle and Lower reaches, respectively.

At the start of the Middle reach, the major part of the discharge is diverted by the weir to the irrigation canal. It is found that the rate of diversion (ratio of yearly diverted to unregulated flow) will be 29%, 32%, and 43% for the Sc-I, Sc-II and Sc-III dam operations, respectively. The dry period river discharges in the Middle (Figure 4.3C) and Lower (Figure 4.3D) reaches will be rather small, since it will represent only the environmental flows.

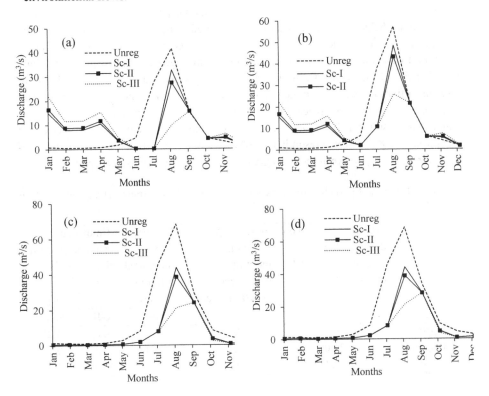

Figure 4.3. Average monthly discharge hydrograph for the four study reaches on the Ribb River for the unregulated (pre-dam) and the three regulated (post-dam) scenarios (Sc-I, Sc-II and Sc-III). (a) Upper-I, (b) Upper-II, (c) Middle and (d) Lower reaches.

## 4.4.2 Pre and Post-Dam Reach-Averaged Sediment Transport Capacity

The monthly reach-averaged pre- and post-dam sediment transport capacity of the river is shown in Table 4.4. Here, the post-dam sediment transport capacity refers to the capacity of the river immediately after dam construction and filling of the reservoir, but before the occurrence of any morphological changes caused by dam construction. The sediment transport capacity will then gradually decrease due to reduced river bed slope and bed material coarsening. Even though the total yearly river discharge volume in the

Upper-I and Upper-II reaches will not change, the sediment transport capacities will present an immediate reduction for all dam operation scenarios. This is due to the reduction of peak discharge values (Figure 4.3A and B). Overall, Sc-II dam operation shows a 53%, 62% and 42% yearly sediment transport capacity reduction in the Upper-I, Upper-II and Middle reaches, respectively, compared to the unregulated (pre-dam) yearly sediment transport capacity of the reaches. It should be noted that the yearly sediment transport capacities computed based on monthly averaged discharges are substantially smaller than those obtained with the daily discharges, see Table 4.4. The sensitivity of sediment transport capacity to the grain size was studied by changing $D_m$ for Sc-II dam operation for the Upper-I reach. The analysis shows that, a reduction of $D_m$ from 7 to 6 mm would result in a 32.7% increase of sediment transport capacity, while an increase of $D_m$ from 7 to 8 mm would result in a 25.6% reduction. The difference is due to the non-linear relationship between sediment grain size and sediment transport capacity.

Table 4.4. Pre- and post-dam reach-averaged yearly sediment transport capacity of Ribb River.

| Reach | Pre-dam $Q_s$ based on daily discharges (Mulatu et al., 2018) ($1 \times 10^4$ m³/year) | Pre-dam $Q_s$ based on monthly discharges ($1 \times 10^4$ m³/year) | Post-dam $Q_s$ (immediately after dam construction) based on monthly discharges for different dam operation scenarios ($1 \times 10^4$ m³/year) | | |
|---|---|---|---|---|---|
| | | | Sc-I | Sc-II | Sc-III |
| Upper-I | 69.5 | 9.35 | 5.17 | 4.4 | 3.25 |
| Upper-II | 20.0 | 2.98 | 1.51 | 1.13 | 0.12 |
| Middle | 6.11 | 4.62 | 2.98 | 2.69 | 1.86 |

## 4.4.3 Results of the Equilibrium Theory

For the gravel-bed reaches, the theoretical equilibrium bed slopes attained by the river a long time after the start of the dam operation are computed with Equation 4.3. In the Upper-I reach, the three dam operation scenarios (Sc-I, II and III) result in longitudinal bed slopes of 0.088%, 0.105% and 0.132%, respectively, where the pre-dam slope being 0.30%. These slopes correspond to theoretical river bed degradations near the toe of the dam of 21.2 m, 19.5 m and 16.8 m, respectively. Final-state theoretical river bed slopes of 0.053%, 0.06% and 0.1% are obtained for the Upper-II reach, where the pre-dam slope being 0.12%. For the upper reaches, the least-impacting scenario is Sc-III, corresponding to the largest water withdrawal for irrigation.

For the sand-bed Middle reach, the post-dam river bed slope is computed using Equation 4.6. In this case, the result is a flat river bed for all dam operation scenarios corresponding to a theoretical bed degradation of 10 m near the base of the weir. This indicates that the analytical method cannot differentiate the least- impacting dam operation scenarios in the sand-bed river reach if it is assumed that all incoming sediment is intercepted by the reservoir and for no sediment input from tributaries. Adopting the Engelund and Hansen (1967) formula implies assuming that the critical value of flow velocity for sediment motion is zero. This means that all dam operation scenarios with zero sediment input to the downstream river ultimately result in a theoretical flat river bed, since the approach did not consider other sediment sources, for instance ephemeral tributaries and bank erosion. Instead, using a sediment transport formula including a threshold for sediment motion, as the one used for the gravel-bed reaches (Wong and Parker, 2006), results in a different final bed slope for each discharge regime. However, for the assumed 20% sediment release to the downstream reaches, the Middle reach may attain an equilibrium bed slope of 0.0017%, 0.0015% and 0.0010% for Sc-I, Sc-II and Sc-III dam operations, respectively.

## 4.4.4 Calibration and Validation of 1D Morphodynamic Model

The results of model calibration show that the measured and simulated water depth values agree quite well for a Chézy coefficient value of 35 $m^{1/2}$/ s resulting in a RMSE value of 0.10 (Figure 4.4a). It is noticed that the model overestimates water levels at low-flows (Figure 4.4b). Model validation also shows a good agreement between the measured and the simulated water depths (Figure 4.4c) with a corresponding RMSE of 0.10.

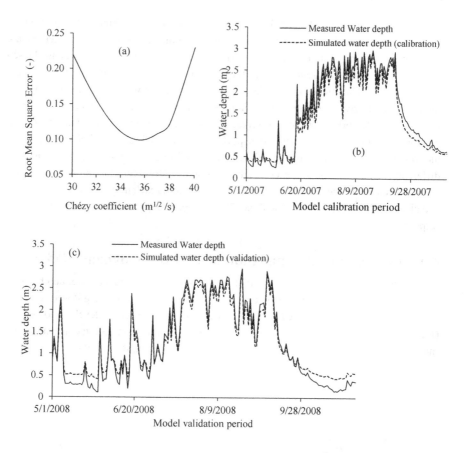

Figure 4.4. (a) Root Mean Square Error for different Chézy coefficient values. (b, c) Measured and simulated water depths at the Lower gauging station for the months of May to October 2007 (Calibration) and May to October 2008 (Validation), respectively.

## 4.4.5 1D Model Application on the Ribb River

The results of the long-term simulation of the calibrated and validated model show two patterns: degradation and aggradation, even though degradation dominates the river system. River bed degradation commences immediately after the start of dam operation at the base of the dam and at the base of the weir propagating in a downstream direction as time passes. At the end of the simulation period, the simulated river bed degradations at the base of the dam is 19, 17 and 12 m for Sc-I, Sc-II and Sc-III, respectively (Figure 4.5a). The simulated river bed degradation at the base of the dam is already 3.4, 2.9 and

2.2 m after 10 years from the start of dam operation for Sc-I, Sc-II and Sc-III, respectively. Bed degradation occurs at rates of 3.8, 3.3 and 2.3 cm/year in the first 5 years and continues at a rate of ~0.1 cm/year at the end of the computational period (Figure 4.5a). Two years after dam closure, simulated bed degradation has extended to 3.2, 3 and 2.6 km from dam site for Sc-I, Sc-II and Sc-III, respectively, whereas it reaches the end of the Upper-I reach (10 km) after 100, 500 and 1,000 years, respectively (Figure 4.5b). At the end of the simulations, the reach-averaged river bed slope of the Upper-I reach is reduced to 0.106%, 0.125% and 0.155% for Sc-I, Sc-II and Sc-III, respectively.

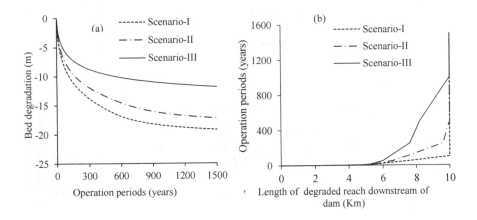

Figure 4.5. Results of the 1D morphodynamic model. (a) Ribb River bed degradation near the base of the dam for different dam operation scenarios over a 1,500 year period and (b) downward propagation of river bed degradation.

In the Upper-II reach, the effects of the dam are felt only after the sediment leaving the Upper-I reach starts to decrease. In the meantime, the Upper-II reach is affected by the weir since sediment movement from upstream to downstream of the weir is zero for the first 40, 60 and 500 years for Sc- I, Sc-II and Sc-III, respectively, until the bed reaches the top of the weir (2.4 increase), Figure 4.6. Simulated bed degradation at the base of the weir is 2.1 m for Sc-I (Figure 4.6a) and 2.3 m for Sc-II (Figure 4.6b) in the first 50 and 80 years of model simulation, respectively. Thereafter, the transported sediment starts to pass over the weir temporally resulting in downstream bed aggradation, creating a new type of disturbance to the downstream reach. This bed aggradation process reduces the rate of morphological changes of the downstream reach. At the end of the simulation period, river bed degradation near the base of the weir is 4.8 m (Figure 4.6a), 5.0 m (Figure 4.6b), and 7.5 m (Figure 4.6c) for Sc-I, Sc-II and Sc-III, respectively. The maximum bed degradation at the base of the weir corresponds to the operation scenario with the highest rate of discharge diversion (Sc-III). Due to upstream river bed degradation, the reach-averaged bed slope of the Middle reach reduces to 0.029% for Sc-I and Sc-II and to 0.023% for Sc-III.

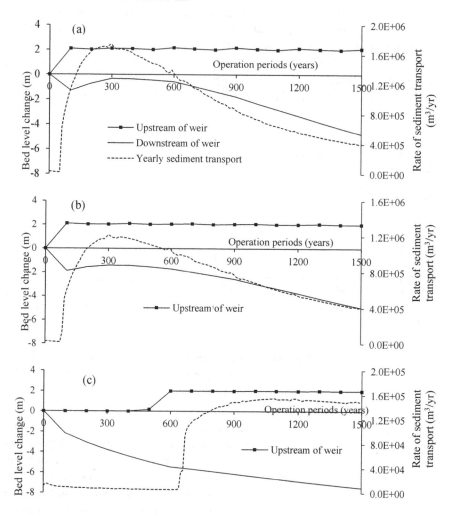

Figure 4.6. Simulated bed level changes immediately upstream and downstream of the Ribb weir, and simulated yearly sediment transport over the Ribb weir since start of the Ribb Dam operation for Sc-I (a), Sc-II (b) and Sc-III (c).

The results of the sensitivity analysis show that a decrease of $D_m$ from 7 to 6 mm would increase the bed degradation at the dam base by 32.5%, whereas an increase of $D_m$ from 7 to 8 mm would decrease the bed degradation by 25.2%. This clearly indicates that the model results are sensitive to the choice of mean sediment grain size.

## 4.4.6 Comparison of Method Results

Here, the results of the ET and 1D morphodynamic model are compared to assess the applicability of the analytical method to distinguish the impact of the different dam operation scenarios and to estimate the final slope of the river reaches. This would also allow to quickly calculate the value of bed degradation at the base of the dam and base of the weir. After 100 years of dam operation, the reach-averaged river bed slope of the Upper-I reach estimated with the 1D model is 51%, 45% and 36% steeper than the one estimated by the ET for the Sc-I, Sc-II and Sc-III, respectively. After 1,500 years, the bed slopes simulated by the 1D morphodynamic model are still 17%, 16% and 15% steeper than those predicted by the ET (Table 4.5). Similarly, the estimated river bed degradation at the base of the dam by the ET exceed that obtained by the 1D model by 2.2, 2.5 and 4.8 m for the Sc-I, Sc-II and Sc-III, respectively.

Table 4.5. Reach-averaged river bed slopes downstream of the Ribb Dam estimated using the Equilibrium Theory and the1D model.

| Reach | Current river bed slope ($i_0$), (%) | Simulation period (years) | Estimated river bed Slope ($i_\infty$), (%) | | | | | |
|-------|------|------|------|------|------|------|------|------|
| | | | 1D model | | | Equilibrium Theory | | |
| | | | Sc-I | Sc-II | Sc-III | Sc-I | Sc-II | Sc-III |
| Upper-I | 0.3 | 100 | 0.178 | 0.191 | 0.206 | 0.088 | 0.105 | 0.132 |
| | | 500 | 0.137 | 0.154 | 0.17 | | | |
| | | 1,000 | 0.113 | 0.134 | 0.159 | | | |
| | | 1,500 | 0.106 | 0.125 | 0.155 | | | |
| Upper-II | 0.12 | 100 | 0.12 | 0.12 | 0.132 | 0.053 | 0.06 | 0.1 |
| | | 500 | 0.108 | 0.112 | 0.132 | | | |
| | | 1,000 | 0.09 | 0.098 | 0.122 | | | |
| | | 1,500 | 0.081 | 0.09 | 0.121 | | | |
| Middle | 0.04 | 100 | 0.036 | 0.033 | 0.033 | 0 | 0 | 0 |
| | | 500 | 0.041 | 0.038 | 0.026 | | | |
| | | 1,000 | 0.035 | 0.034 | 0.025 | | | |
| | | 1,500 | 0.029 | 0.029 | 0.023 | | | |

For the assumed 20% sediment release from the upper reaches the ET calculated a gentler slope for the Middle reach than the 1D model for all dam operation scenarios. At the end of the 1D model simulation period, bed degradation at the base of the weir still occurs at rates ranging from 0.2 (Sc-III) to 0.46 cm/year (Sc-I), indicating an on-going morphological trend after 1,500 years. In general, the ET overestimates the morphological changes for both gravel-bed and sand-bed reaches compared to the 1D model for the considered period of model simulation, as shown in Table 4.5. With the discussed limitations of the analytical and 1D morphodynamic model approaches, here the advantages gained and the easiness of the analytical approach compared with the 1D morphodynamic modelling (Table 4.6) was presented. The comparison items listed in the table are not comprehensive, but assumed enough for discussion. From the comparison of results, it is possible to say that the analytical models yields acceptable results for the comparison of different dam operation scenarios in the preliminary level of study for selecting the dam site.

Table 4.6. Comparison of analytical and 1D numerical models.

| Comparison items | Analytical approach (ET method) | 1D modelling approach (SOBEK-RE) |
|---|---|---|
| Cost related to required input data | Can be applied for the reach-averaged river geometry. | Detail river channel cross-sectional surveying works or high resolution satellite images are required to extract river channel geometry. For model calibration and validation, historical data related to pre- and post-damming is required, which are lacking in the system. |
| Cost related to modelling experience | Very simple to apply at field level to compare different sites and dam operation scenarios. | Modelling background and experience is required, which incur costs to train experts. |
| Time for analysis | Results can be obtained within hours. | Results can be obtained after several days/weeks of modelling. Before simulation, substantial amount of time is required for model calibration and validation. |

## 4.5 DISCUSSION

Geologically, the Upper Blue Nile Basin is formed by a combination of tecnto-volcanic activities, and quaternary superficial processes (erosion and deposition) dominated by Oligocene-Miocene volcanic formations overlain by Quaternary alluvio-lacustrine deposits on the floodplain (Chorowicz et al., 1998). As described by Poppe et al. (2013) and Chorowicz et al. (1998), the basin shows gradual geological changes. This means that the long-term evolution of longitudinal river bed slope may not be affected by the geological setting and their dynamics (uplift, rifting and subsidence) and tectonic condition of the area. On the whole, the existence of disintegrated rock outcrops at some locations in the Upper-I and Upper-II reaches, sediment supply from ephemeral tributaries (Benda et al., 2004), bed material coarsening and armoring (Williams and Wolman, 1984; Kondolf, 1997) may limit the simulated depth of degradations and reach averaged slopes.

In gravel-dominated river reaches, post-dam sediment transport occurs only at discharges exceeding the limits for sediment motion (Gaeuman et al., 2005; Grant, 2012). Application of Wong and Parker (2006) sediment transport capacity formula for the current river conditions (Section 4.4.2) shows that the bed material in the Upper-I and Upper-II reaches does not move if the discharge is less than 10 and 21.5 m$^3$/s, respectively. The reduced peak discharge during the wet season may not have the capacity to transport the largest sediment particles, thereby forming an armor bed surface (Kondolf, 1997; Grant, 2012), which could shorten the period to attain equilibrium. For the sand dominated reaches, the analysis shows that there may be some sediment movement for the entire range of flow releases until the river bed becomes completely flat for no sediment input from the upper reaches. However, the reaches may attain an equilibrium at a certain slope greater than the theoretical one as there may be sediment supplied during the rainy season from the ephemeral tributaries. Generally, the morphological activities in the sand bed reaches dominate the morphological time scale of the Ribb River system. Similar results are also found for the Middle Yangtze River, China (Zhou et al., 2018). The overall reduction of flow in this reach may also create favorable conditions for the growth of vegetation in parts of the old channel which will reduce the channel width (Williams and Wolman, 1984; Schmidt and Wilcock, 2008).

The 1D model and the ET show that the dam operation scenario that yields the maximum dry- or wet-period peak flows (Sc-I) produces the maximum bed degradation downstream of the Ribb Dam, the minimum longitudinal bed slope in the upper reaches and the fastest downward propagation of bed degradation (Figure 4.5 and Figure 4.6b). On the other hand, this scenario produces a relatively small morphological impact on the reach downstream of the weir, as some decades are required to replenish the degraded channel by the overpassed sediment (Figure 4.6a). It is important to note that a large portion of the total bed degradation at the dam base, which requires remedial measures, is obtained in the first 100 years, namely 58% for Sc-I and 56% for Sc-II and Sc-III dam operations.

The application of the ET, which assumes that the study reaches are fully alluvial, resulted in the assessments of the theoretical ultimate river bed slopes after the start of dam operation. These slopes are attained when no sediment deposition and no bed erosion take place in the considered reaches (new equilibrium). The results of this study show that the ET overestimates the morphological changes compared to the 1D model (after 1,500 years). Duró et al. (2016) found a good agreement between the slopes predicted by a 2D Delft3D morphological model and the ones derived with the ET by analysing the long-term morphological effects of channel width variations; though the ET tended to slightly overestimate the changes. The ET method can only be used to compare dam operation scenarios on the sand-dominated reaches if there is sediment supply from the upper reaches.

Li et al. (2014) suggested that a river system may be considered in equilibrium if the maximum bed level changes are less than 0.5 cm/year at the upstream boundary. The analysis of the 1D model result shows that the bed level change rates are reduced to 0.5 cm/year after 700 years for Sc-I and Sc-II and, after 400 years for Sc-III. De Vries (1975) derived the following formula for the assessment of the time scale of the morphological adaptation of entire river reaches (see also Jansen et al. (1979)):

$$T = \frac{3BL^2 i(1-\varepsilon)}{bQ_s} \tag{4.12}$$

in which T is the morphological adaptation time when the morphological change has attained 50% of the total.

De Vries (1975) estimated the 50% morphological change adaptation time of the Dutch Rhine River reach, assumed 200 km long, and found it to be 1,000 years. Similarly, Church (1995) analyzed the time scale for 50% morphological change adaptation of the Peace River, located in British Columbia and Alberta, to the operation of the WAC Bennett Dam, using the approach of De Vries (1975) and found it to be greater than 10,000 years for the upstream cobble dominated reach and 5,000 years for the sand dominated downstream reach. Di Silvio and Nones (2014) also found similar results, in which larger rivers may require hundreds and thousands of years to attain a final equilibrium after disturbance. However, it was not possible to apply the same method to the entire Ribb River, from the dam site to the Lake Tana, because the weir and the water extraction for irrigation subdivide the river in two distinct parts, with different discharge regimes. Applying the method to the two parts of the river separately would strongly underestimate the adaptation time, considering that the morphological adaptation of the river reaches downstream of the weir are influenced by the morphological changes in the upper reaches, which affect the amounts of sediment bypassing the weir.

Both methods used in the study were based on simplifications that might affect their results, as, for instance, the use of reach-averaged channel geometries and granulometric characteristics. Furthermore, both methods did not include the sediment input from bank erosion and ephemeral tributaries. This introduces uncertainty into the computed results. Consequently, the river bed slopes obtained by both the ET and by the 1D morphodynamic model should be interpreted as mere indications of the morphological adjustment, considering also that the river may never achieve equilibrium as the forcing and boundary conditions such as the river discharge, sediment yield, population density, land use, land cover, etc., fluctuate through time (Bolla Pittaluga et al., 2014). Nevertheless, both approaches used in this study allow discriminating the effects of different dam operation scenarios on gravel and sand-bed rivers. However, the application of the ET on sand-bed rivers is restricted to the cases in which some sediment is supplied from the upper reaches. Validation of both methods to determine if they indeed allow recognizing the least impacting scenario is difficult, considering that data showing the effects of different scenarios on the long-term river morphology are lacking.

## 4.6 CONCLUSIONS

This study addresses the applicability of the physics-based analytical equations (Equilibrium Theory) compared to a 1D numerical model (SOBEK-RE) to determine the least-morphologically impactful dam operation scenario on the river reaches downstream of the under-construction Ribb River dam, which are dominated by gravel and sand. Three scenarios were evaluated: Sc-I, corresponding to a dam operation with a command area 10% smaller than the design command area; Sc-II, corresponding to a dam operation for the design command area; and Sc-III, corresponding to a dam operation with a command area 32.8% greater than the design command area.

Both methods show that Sc-I has the greatest morphological impact immediately downstream of the Ribb Dam, as it has the smallest reduction in peak discharge values (Figure 4.5a). The 1D model simulation showed a bed degradation at the dam base of 19 m after 1,500 years, whereas a bed degradation of 21.2 m was estimated by the ET. Bed degradation starts at the dam site and propagates downstream through time. Upstream of the weir the river bed gradually rises until it reaches the weir crest level. Immediately downstream of the weir, the river bed shows first degradation, then aggradation, starting when the upstream bed level has reached the crest, and then again degradation, starting when the sediment input from the reach Upper-II reduces to a value below the sediment transport capacity of the flow in the Middle reach. This type of alternating bed level changes at the base of the weir slows down the morphological adaptation of the downstream sand-bed reaches considerably (Figure 4.6b).

The ET overestimates the morphological changes for both the gravel-bed and the sand-bed reaches compared to the 1D morphodynamic model for the specified period of model simulation, though both methods show relatively good agreement for the reach close to the dam (Upper-I). This is due to the over-simplification of the theory compared with the 1D model. In addition, the 1D model results show the existence of morphological activity in all the reaches even after 1,500 years for all dam operation scenarios (Section 4.4.5). Generally, the model comparison indicates that the ET can be used to roughly and quickly assess the morphological changes in river reaches downstream of a dam for feasibility studies, considering that the method tends to overestimate morphologic adjustment.

Like dams, the long-term operation of diversion weirs will affect the fractional movement of sediments (Jansen et al., 1979; Thoms and Walker, 1993; Ahn et al., 2017). Application

73

of the ET for the sand dominated river reaches downstream of the weir results in ultimate flat river bed, with a theoretical bed degradation of 10 m at the base of the weir for all dam operation scenarios for no additional sediment input either from the tributaries or released from the dam. However, the model can be used for detecting the least impacting dam operation scenario for sand dominated reaches if there is sediment supply from the upper reaches. The 1D model simulation for the weir downstream reaches show that the morphological equilibrium will not be attained before several centuries.

# 5

# ALTERATION OF THE FOGERA PLAIN FLOOD REGIME DUE TO RIBB DAM CONSTRUCTION[3]

The Fogera Plain, located on the eastern shore of Lake Tana is affected by recurrent flooding by the Ribb and Gumara Rivers. This chapter analyses the effects of the Ribb Dam, currently under construction, on extension and duration of flooding, focusing on the wetlands in the Fogera plain which provide important habitats for fish and bird species.

---

[3] This chapter is based on:
Mulatu, C. A., Crosato, A., M., Langendoen, E. J., Moges, M., & McClain, M. (2021). Alteration of the Fogera Plain flood regime due to Ribb Dam construction, Upper Blue Nile Basin, Ethiopia. *Journal of Applied Water Engineering and Research*. doi.org/10.1080/23249676.2021.1961618.

## 5.1  INTRODUCTION

Flooding is a natural phenomenon that occurs when river natural and manmade drainage channels do not have enough capacity to handle the discharge generated from their watersheds (Leon et al., 2014; Teng et al., 2017). It affects infrastructures like roads, schools, houses, water points, etc., and has caused the death of millions of peoples worldwide. Among the so-called natural disasters, flooding caused more than half of fatalities (Berz, 2000; Ohl and Tapsell, 2000; Opperman et al., 2009), and it accounts for one-third of the economic losses (Pilon, 2002). In recent decades, the frequency of floods and associated damages have increased rapidly due to the increase in population and economic activities in fertile flood-prone areas and climate changes (Svetlana et al., 2015). According to the international disaster database report, cited by Hu et al. (2019), on average, 85 million people were affected per year between 2007 and 2016, with an annual economic loss of 36.7 billion US dollars. It is believed that flood disasters cannot be avoided; however, the associated impacts can be reduced by awareness and preparedness. Floods also provide benefits for the ecology of floodplains, rivers, wetlands, and estuaries (FitzHugh and Vogel, 2011). This includes the replenishment of soil nutrients, promoting the upriver migration of fishes for spawning, and support of plant and aquatic habitats.

Dams reduce flooding by attenuating peak discharge values (Petts, 1980; Kondolf, 1997; Petts and Gurnell, 2005; Graf, 2006; Mei et al., 2017). The immediate hydrological effects are a change in frequency of high and low flows and their time of occurrence, reducing the peak and increasing the low flows to produce a new hydrograph (Williams and Wolman, 1984; Magilligan and Nislow, 2005; Graf, 2006; Ronco et al., 2010; Grant, 2012). However, the extent of attenuation will vary with the rate of runoff and the ratio between peak inflow and outflow (Kondolf, 1997). These changes affect the flooding extent of downstream floodplains and lowers the groundwater table, which may adversely impact habitats needed for the survival of endemic floodplain species (Talukdar and Pal, 2019; Wild et al., 2019; Li et al., 2020).

Peak discharge attenuation promotes conditions for vegetation encroachment and growth on river banks, which controls the river width by reducing bank erosion. Discharge attenuation affects downstream wetlands and agricultural activity (Kondolf, 1997) by decreasing lateral connectivity between floodplains and river channels, which is important to maintain wetland habitats in the floodplain (Ward et al., 2002; Julian et al., 2016; Talukdar and Pal, 2019). The blockage of sediment and debris to the downstream floodplain also affects soil fertility, habitat complexity and reduces food for aquatic species (Qicai, 2011).

The replenishment of fertile, clay soils by recurrent flooding of the Ribb and Gumara Rivers makes the vast agricultural land of the Fogera Plain suitable for enhanced agricultural production. However, the flooding often results in fatalities and displaced people, which adversely affects the local economy. A dam is under construction in the headwaters of the Ribb River to impound 234 million m$^3$ of water that will be used to reduce flooding and irrigate 15,000 ha of command area (BRLi and MCE, 2010). The construction of the Ribb irrigation system will affect the direct supply of water and nutrients from the river to the wetlands in the Fogera Plain. Hence, the floodplain may face important changes in the near future, which on one hand are expected to improve the local economy and safety, but on the other hand, may negatively impact the floodplain ecosystem.

The main objective of this study is to analyze the potential effects of Ribb Dam operation on the river discharge regime and the extent and duration of flooding in the Fogera Plain. It also assesses the potential implications of hydrological alteration on the floodplain ecology of the wetlands. The HEC-HMS (Hydrologic Engineering Center-Hydrologic Modeling System) hydrological model (version 4.3) was applied to determine the discharge time-series of the major rivers entering the Fogera Plain, as the existing discharge measurement stations are unable to gauge the overbank flow portion, and for the Ribb reservoir flood routing. The model output was used as input for an HEC-RAS 2D (Hydrologic Engineering Center-River Analysis System) hydrodynamic model (version 5.0.7) to compute the pre- and post-dam extent of flooding in the Fogera Plain. Historical inundation maps were retrieved for 01 August 2010 from the spectral difference of the ground objects reflectance values of Landsat satellite images using the cloud-computing platform of Google Earth Engine to calibrate the hydrodynamic model. Even though the approach in this study does not present a novel understanding of the effect of dams on downstream flooding and ecological changes, it provides important new knowledge to apply specifically in the Upper Blue Nile Basin. Moreover, the approach and findings of the work can also be used as a baseline to study similar river basin planning efforts, especially in ungauged and scarcely gauged river basins.

## 5.2 DESCRIPTION OF THE STUDY AREA

The Fogera Plain is located on the eastern periphery of Lake Tana, the source of the Blue Nile River, Ethiopia. The Ribb and the Gumara are the two major perennial rivers that pass through the Fogera Plain and cause flooding (SMEC, 2008b; Mulatu et al., 2018). The rivers originate in the Guna Mountains and flow westward receiving additional discharge from tributaries to drain to Lake Tana (Figure 2.6). The ungauged watersheds surrounding the Ribb and Gumara Rivers, which are either drained by ephemeral tributaries or by overland flow also contribute discharge to the floodplain. The central part of the ungauged watershed (Figure 2.6) has an area of 200 km$^2$, in which 75.7% and 10% of the area are dominated by cultivation and farm villages, respectively. This watershed does not have a defined drainage channel but drains directly to the Fogera Plain by small gullies. 64% of the Ribb and Gumara watersheds are dominated by cultivation, while 11% of the Gumara and 10% of the Ribb watersheds are covered by shrubs and bushlands. The watersheds are dominated by a tropical climatic condition: humid and warm in the rainy and dry seasons, respectively. Near the Lake Tana shore, the average monthly maximum temperature varies from 17 °C to 37 °C and from 19 °C to 39 °C for the Gumara and Ribb watersheds based on data of Woreta and Addis Zemen meteorological stations (1995 to 2015), respectively. In December, the temperature in the headwaters of the Gumara and Ribb watersheds falls below 0 °C (Debre Tabor meteorological station). The Ribb and Gumara watersheds receive an average yearly rainfall of 1,300 mm and 1,320 mm, respectively, with higher precipitation in their upper mountainous areas and less near the lake shore. The rainfall is unimodal and 80% falls in the rainy season (June to September).

## 5.3 DATA SOURCES

The data required for the study were obtained from different sources (Table 5.1). This includes the collection of the time-series discharge data of the Ribb and Gumara Rivers and Lake Tana water levels from the Ethiopian Ministry of Water, Irrigation, and Electricity (MoWIE) and, the rainfall data at the nearby meteorological stations from the Ethiopian National Meteorological Agency (NMA). Literature such as SMEC (2008b), Dessie et al. (2014), Abate et al. (2015), Mulatu et al. (2018) and Mulatu et al. (2020) were reviewed to understand the current state of river discharge, anthropogenic effects, and geometric characteristics. The data needed for the Ribb reservoir flood routing were extracted from the Ribb Dam feasibility study and design documents (WWDSE and TAHAL, 2007).

Table 5.1. Collected data and their sources. TM for Landsat 5 satellite image stands for Thematic Mapper indicating the type of earth observing instrument mounted on the satellites.

| Data type | Data period | Resolution | Source |
|---|---|---|---|
| Digital Elevation Model (DEM) | 2014 | 30 m by 30 m | https://earthexplorer.usgs.gov/ |
| Ribb and Gumara River discharges | 2007-2010 | Daily | Ministry of Water, Irrigation, and Electricity of Ethiopia (MoWIE) |
| Ribb and Gumara River cross-sections | ------ | ------- | MoWIE (collected during the study and design of Ribb irrigation project) |
| Lake Tana water level | 2007-2019 | Daily | MoWIE (Bahir Dar station) |
| Rainfall at the nearby meteorological stations | 2007-2019 | Daily | Ethiopian National Meteorological Agency |
| Landsat 5 satellite images (L5-TM) | 2010 and 2011 | 30 m by 30 m | http://landsat.gsfc.nasa.gov/ |

## 5.4 METHODOLOGY

This study applied a combination of hydrologic and hydrodynamic computer models to investigate the effectiveness of dam construction at the upper reach in reducing flooding in the Fogera Plain. The study also relates the alteration and/or reduction of discharge due to operating the Ribb Dam and the irrigation system to the ecological dynamics of wetlands on the Fogera Plain. The integration of hydrologic and hydrodynamic models and the general workflow are shown in Figure 5.1. Flood inundation maps were derived from the spectral difference of ground object reflectance values of Landsat satellite images. The hydrologic model is presented in Section 5.4.1. The peak discharge attenuation due to dam construction and the time series outflow discharge from the reservoir are assessed in Section 5.4.2. The method to determine historical open water surface areas from the 30 m resolution of Landsat satellite images using the Google Earth Engine (GEE), a cloud computing platform, is discussed in Section 5.4.3. The hydrodynamic model to examine flooding of the Fogera Plain and the ecological dynamics of the wetlands are presented in Section 5.4.4 and Section 5.4.5 respectively.

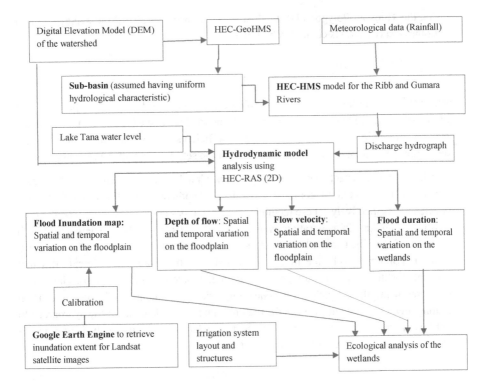

Figure 5.1. Work Flow and model integration.

## 5.4.1 Pre-dam Discharge Analysis

The Lower Gauging stations of the Ribb and the Gumara Rivers are located within the floodplain where the peak discharges that cause flooding are not confined by the banks (SMEC, 2008b; Dessie et al., 2014; Mulatu et al., 2018). For example, Dessie et al. (2014) indicated a peak flood reduction, relative to the rated discharge, of the Ribb River at the lower gauging station up to 71% between December 2011 and December 2012. Therefore, the discharge data of the rivers derived from gauging stations' rating curves cannot be used for flood flow analysis in this study. Instead, a HEC-HMS rainfall-runoff model was developed for the Ribb and Gumara River watersheds to determine the daily time-series discharge data. HEC-HMS is developed by the United States Army Corps of Engineers (USACE) and simulates the rainfall-runoff process of dendritic watershed systems (Scharffenberg and Fleming, 2016). It is selected as it has been successfully applied to simulate both short- and long-term rainfall-runoff events in Ethiopia (Halwatura and Najim, 2013; Demlie G Zelelew and Assefa M Melesse, 2018; Tassew et al., 2019). Moreover, data availability, the ability of the model to operate huge tasks with hydrological studies, and modeling experience were considered.

Using the Thiessen Polygon (TP) method (e.g., Shaw et al. (2010)) the average rainfall over the area ($RF_{aerial}$) is computed by Equation 5.1.

$$RF_{aerial} = \sum_{i=1}^{n} \left( \frac{R_i * A_i}{A_t} \right) \tag{5.1}$$

where $R_i$ is the rainfall at the i-th meteorological station (mm/day), $A_i$ is the polygon area of the i-th meteorological station (km$^2$), $A_t$ is the total watershed area (km$^2$), and n is the number of stations. Once the Thiessen coefficients ($A_i/A_t$) for the stations were determined, the areal rainfall was computed for the set of rainfall measurements.

The initial and constant method and the Clark Unit Hydrograph (UH) transformation method were selected to compute the rainfall depth and the direct runoff for the watersheds, respectively. These methods are selected based on data availability and used mostly to simulate short-duration events (Jin et al., 2015; Demlie Zelelew and Assefa Melesse, 2018). The Clark UH method requires the time of concentration ($T_c$, hr), which is the maximum travel time in the watershed, and the storage coefficient (Ra) that accounts for the temporary storage of excess precipitation (Feldman, 2000). The $T_c$ was estimated using the Kirpich (1940) equation, written as Equation 5.2. The lag channel routing method, which determines the required time to translate the inflow hydrograph to outflow without attenuation, is selected as it is widely used for the analysis of drainage channels (Scharffenberg and Fleming, 2016).

$$T_c = 0.00013 * L_m^{0.77} * S_L^{-0.385} \tag{5.2}$$

where $L_m$ is the maximum flow length (ft) and $S_L$ is the channel slope (-).

The HEC-HMS hydrological model was calibrated and validated against the daily time-series discharge data of the Lower Gauging stations for 'dry-years' in which the river discharge does not breach the banks (Table 5.2). The availability of continuous discharge and rainfall data for the respective hydrological and meteorological stations were also considered.

Table 5.2. HEC-HMS model calibration and validation periods for the Ribb and Gumara watersheds.

| Name of Watershed | Model calibration | | Model validation | |
|---|---|---|---|---|
| | From | To | From | To |
| Ribb | 1-Jun-2007 | 30-Sep-2007 | 1-Jun-2008 | 30-Sep-2008 |
| Gumara | 1-Jun-2008 | 30-Sep-2008 | 1-Jun-2009 | 30-Sep-2009 |

The model performance was measured using the Nash-Sutcliffe Efficiency (NSE) and coefficient of determination ($R^2$) methods written as Equation 5.3 and 5.4, respectively. Also, the comparison between the simulated and observed volume and peak discharge values and date of occurrences were considered. Watershed (the constant rate and initial loss scale factors, and the time of concentration and the storage coefficient transform methods) and reach (the lag time) parameters were automatically optimized during the calibration process by maximizing the NSE and $R^2$ values.

$$NSE = 1 - \frac{\sum_{i=1}^{n}\left(Q_{m,i} - Q_{s,i}\right)^2}{\sum_{i=1}^{n}\left(Q_{m,i} - \bar{Q}_m\right)^2} \tag{5.3}$$

$$R^2 = \frac{\left[\sum_{i=1}^{n}\left(Q_{m,i} - \bar{Q}_m\right)\left(Q_{s,i} - \bar{Q}_s\right)\right]^2}{\sum_{i=1}^{n}\left(Q_{m,i} - \bar{Q}_m\right)^2 \sum_{i=1}^{n}\left(Q_{s,i} - \bar{Q}_s\right)^2} \tag{5.4}$$

where $Q_{m,i}$ and $Q_{s,i}$ are the measured and simulated daily discharge values (m$^3$/s), respectively, and $\bar{Q}_m$ and $\bar{Q}_s$ are the measured and simulated average daily discharge values (m$^3$/s), respectively. Model performance can be considered very good, good, satisfactory, and unsatisfactory if $R^2$ and NSE vary from 0.75 to 1, 0.65 to 0.75, 0.50 to 0.65, and < 0.5, respectively (Moriasi et al., 2007).

The calibrated and validated watershed model for the dry-years (Table 5.2) was run from 2010 to 2019 to generate daily pre-dam river discharge time-series data for the Ribb and Gumara Rivers at their Lower Gauging stations including peak discharge values. Sub-watersheds, which are an important input parameter for the HEC-HMS model (Figure 5.2) were generated from the DEM of the watershed using HEC-GeoHMS, an ArcGIS extension.

The sensitivity of the peak discharge and discharge volume values to the watershed and reach parameters was determined by varying the parameters between -30% and +30% with respect to the calibrated values. Sensitivity analysis informs how the model output is affected by changing the model simulation parameters (Rauf and Ghumman, 2018). Moreover, watershed and reach parameter uncertainty analysis was done using a Monte Carlo analysis of HEC-HMS 4.3 manager for 1000 trials. As described by Oubennaceur et al. (2018) and Shamsudin et al. (2011), this is used to provide an insight into how well the model is calibrated and to identify the possible range of model parameters.

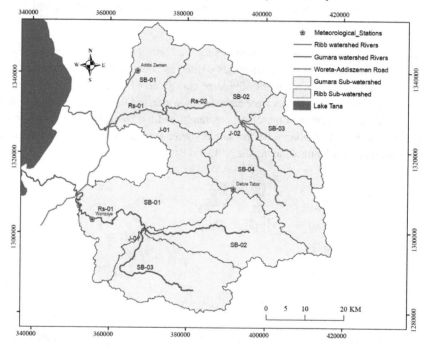

Figure 5.2. Watersheds for Ribb and Gumara Rivers and reaches (Rs) with river networks with nearby meteorological stations. The abbreviations SB and J stand for sub-watersheds and the junctions of the reaches, respectively.

## 5.4.2 Post-dam Discharge Analysis

The Ribb Reservoir will impound 234 million m$^3$ of water and inundate an area of 10 km$^2$ at the Normal Pool Level (NPL) elevation of 1,940 m a.s.l. (WWDSE and TAHAL, 2007).

Sufficient runoff is produced in the rainy season to fill the reservoir fairly quickly; the excess water will pass through the spillway. However, due to the large surface area of the reservoir, the incoming instantaneous peak discharge will be attenuated during outflow.

The HEC-HMS model was developed to determine the time-series outflow discharge using the Modified Puls flood routing method that assumes the reservoir water surface is horizontal. The basic relation for the reservoir flood routing is the continuity of flow, which states that during any time interval the inflow volume is equal to the outflow volume plus/minus change in storage (Fenton, 1992; Yang and Cai, 2011), given by Equation 5.5.

$$I_{avg} - Q_{avg} = \frac{\Delta S}{\Delta t} \tag{5.5}$$

where $I_{avg}$ and $O_{avg}$ are the average daily inflow and outflow discharges (m$^3$/s), respectively, over a daily time step $\Delta t$ (=86400 s), and $\Delta S = (S_{t+1} - S_t)$ is the daily change in storage (m$^3$). HEC-HMS approximates Equation (5.5) as Equation (5.6).

$$\left(\frac{2S_{t+1}}{\Delta t} + O_{t+1}\right) = (I_t + I_{t+1}) + \left(\frac{2S_t}{\Delta t} - O_t\right) \tag{5.6}$$

where the subscripts t and t+1 refer to variable values at the beginning and end of a time step, respectively. To solve for $O_{t+1}$, the reservoir Elevation-Area-Discharge and the ogee-shaped spillway equations were selected to relate storage to outflow discharge at time t+1.

The time-series discharge data of the Ribb River at the dam site, which is one of the inputs for reservoir flood routing, is not available for the study period. Hence, the drainage area proportion method was used to calculate the time-series discharge data at the dam site from the simulated pre-dam values of the Lower Gauging station from 2010 to 2019 (Section 5.4.1), assuming the watershed receives uniform rainfall and generates uniform unit discharge. Other modeling inputs, such as irrigation water requirement, average environmental outflow, elevation-storage characteristics of the reservoir, spillway crest elevation, crest length, and approach depth were obtained from the Ribb Dam feasibility study and design documents of WWDSE and TAHAL (2007). The post-dam discharge at the flood starting point of the Ribb River (2010 to 2019) was obtained by combining the dam outflow discharge and the discharge from the watershed downstream of the dam.

## 5.4.3 Pre-dam Flooding

Application of remote sensing techniques to determine open water surface area have increased over the last decade (Teng et al., 2017), even though the precise estimation remains a challenge due to topography, atmosphere, land cover, and sensor limitation (Donchyts et al., 2016). Most of the radiation beyond and inside the near-infrared wavelengths is absorbed by water, enabling the detection of surface water using reflectance spectral index (Donchyts et al., 2016). The Modified Normalized Difference Water Index (MNDWI) (Equation 5.7), which removes the reflectance values of built-up features (Xu, 2006; Ji et al., 2009) was used to determine the open water surface area for the Fogera Plain. The method is applied largely for data-poor regions (Komi et al., 2017; Teng et al., 2017).

$$MNDWI = \frac{Green - SWIR1}{Green + SWIR1} \tag{5.7}$$

where Green and SWIR1 represents the surface reflectance values for the green and the shortwave infrared wavelengths, respectively. The MNDWI value was calculated using surface reflectance values obtained from Landsat imagery collections within the Google Earth Engine (GEE) and it ranges between -1 and 1.

The determination of open water surface extent by GEE is mainly dependent on the selection of threshold reflectance values (Ji et al., 2009; Donchyts et al., 2016; Tang et al., 2016), which vary with image acquisition time and location (Ji et al., 2009; Feyisa et al., 2014). The Otsu (1979) method was used to determine the threshold reflectance value from the MNDWI histogram for the selected time interval of the study area. For MNDWI values that range from X to Z, the Otsu method divides these values into a water class $(X, ..., T)$ and a non-water class $(T, ..., Z)$, where $-1 \leq X \leq T \leq Z \leq 1$ and T is the threshold value (Li et al., 2013). The method employs statistical analysis to determine the probability of the pixel being in the water $(p_w)$ and the non-water $(p_{nw})$ classes. The optimal threshold value is obtained by the in-between class variance of $p_w$ and $p_{nw}$ classes (Li et al., 2013), which is given by Equation 5.8.

$$T = \arg\max_{X \leq T \leq Z} \left\{ p_w \left( M_w - M \right)^2 + p_{nw} \left( M_{nw} - M \right)^2 \right\} \tag{5.8}$$

where M is the mean of MNDWI, and $M_w$ and $M_{nw}$ are the mean MNDWI values for the water and non-water classes, respectively.

The methodology was applied to a cloud-free Landsat 5 satellite image for 01 August 2010, as (1) it represents well the flooding period of the study area and (2) the effect of dam construction on the nature of river flow and Fogera Plain flooding is assumed minimum as it was during the starting phase of dam construction. The obtained inundation extent was used as a base-map to calibrate the HEC-RAS 2D hydrodynamic model. Furthermore, the spatial and temporal variation of inundation over the floodplain was studied for the year 2011, in which cloud-free satellite images were available for 26 June, 19 July, 20 August, 21 September, and 23 October.

## 5.4.4 HEC-RAS 2D Modeling for the Pre-and Post-dam Fogera Plain Flooding

The recent advancement of computer technology has led to the development of different one-dimensional (1D), two-dimensional (2D), hybrid (1D2D) and three-dimensional (3D) mathematical models even though accurate and rapid modeling remains lacking due to the complex nature of the flow and modeling uncertainties (Teng et al., 2017). Some of the leading mathematical models include MIKE FLOOD (Patro et al., 2009; Kadam and Sen, 2012), SOBEK-RURAL (https://www.deltares.nl/en/software/sobek/), HEC-RAS (Brunner, 2016), and LISFLOOD-FP (Bates et al., 2013; Coulthard et al., 2013). For this study, the freely available HEC-RAS 2D (version 5.0.7), developed by the USACE, is selected based on data availability, nature of the problem and, the study objectives. The model was applied successfully for flood studies (Knebl et al., 2005; Mohammadi et al., 2014; Quirogaa et al., 2016) and the 1D version shows good predictive power if calibrated against inundated area or discharge (Horritt and Bates, 2002). Quirogaa et al. (2016) also satisfactorily calibrated the 2D version of the model for the historical inundation area. The model can simulate steady and unsteady flow for supercritical, subcritical, or mixed flow conditions using the 2D Saint Venant or the 2D diffusion wave equations (Brunner, 2016). The 2D Saint Venant equations combines the continuity (Equation 5.9) and the momentum balance equations in the x (Equation 5.10) and y (Equation 5.11) directions.

$$\frac{\partial H}{\partial t} + \frac{\partial (hu)}{\partial x} + \frac{\partial (hv)}{\partial y} + q = 0 \tag{5.9}$$

$$\frac{\partial u}{\partial t} + u\frac{\partial u}{\partial x} + v\frac{\partial u}{\partial y} = -g\frac{\partial H}{\partial x} + v_t\left(\frac{\partial^2 u}{\partial x^2} + \frac{\partial^2 u}{\partial y^2}\right) - c_f u + fv \tag{5.10}$$

$$\frac{\partial v}{\partial t} + u\frac{\partial v}{\partial x} + v\frac{\partial v}{\partial y} = -g\frac{\partial H}{\partial y} + v_t\left(\frac{\partial^2 v}{\partial x^2} + \frac{\partial^2 v}{\partial y^2}\right) - c_f v - fu \tag{5.11}$$

where u and v are the depth-averaged velocities in x and y directions (m/s), respectively, q is the source/sink flux term (m/s), H is the water surface elevation (m), h is the flow depth (m), g is the acceleration due to gravity (m/s$^2$), $v_t$ is the horizontal eddy viscosity coefficient (m$^2$/s), $c_f$ is the bottom friction coefficient (1/s), given as $\left(\left(\eta^2 g|V|\right)/R^{4/3}\right)$, f is the Coriolis parameter (1/s), |V| is velocity magnitude (m/s), R is the hydraulic radius (m), and η is the Manning's roughness coefficient (s/m$^{1/3}$).

The diffusive wave equations are obtained when assuming the inertial terms are negligible compared to the friction and pressure terms in Equation 5.10 and Equation 5.11. The above set of equations is then reduced to Equation 5.12.

$$\frac{\partial H}{\partial t} + \nabla.\beta \nabla H + q = 0 \tag{5.12}$$

where $\beta = \left( -\left( R(H) \right)^{2/3} / n \right)$ and $\nabla$ is the differential operator $\left( \partial/\partial x, \partial/\partial y \right)$.

The stability of the diffusive wave model can be achieved by selecting a time step that satisfies the Courant condition (Equation 5.13).

$$C_r = V_w \frac{\Delta t}{\Delta x} \leq 2.0 \qquad (5.13)$$

where $C_r$ is the Courant Number with a maximum value for model stability of 2 (-), $V_w$ is the flood wave velocity (m/s), $\Delta t$ is the computational time step (s), and $\Delta x$ is the average cell size (m). The model allows using varying time steps for the range of Courant Number conditions to give more stability and faster computation time (Brunner, 2016).

## Model setup

A high-resolution DEM of the study area, showing the river channel details, one of the main input parameters for the model, was lacking. A separate DEM for the river channels was developed using the reach-averaged river cross-sections and this was combined with the existing DEM using HEC-RAS mapper to produce an improved DEM that includes the river channels. Similar procedure was followed for the major roads.

Flood inundation analysis using the hydrodynamic model has a certain degree of uncertainty related to the quality of the time-series discharge data, DEM resolution, and model simulation techniques (Pappenberger et al., 2005; Merwade et al., 2008; Oubennaceur et al., 2018; Pinos and Timbe, 2019). Sayama et al. (2012) and Komi et al. (2017), indicated that uncertainty analysis that combines hydrologic and hydrodynamic modeling is computationally expensive as it requires the analysis of many uncertain variables. Moreover, our study mainly focuses on the comparison of the pre- and post-dam inundation parameters, for which both models may have similar uncertainty as they are developed for similar boundary conditions and simulation parameters. Hence, this study is limited to the sensitivity analysis of model parameters such as grid spacing, computational time step, and equations to understand the behavioral change of the model output for different values of model parameters.

HEC-RAS allows using large grid spacing as the model extracts detailed elevation values of the cell face like a cross-section from the underlying DEM (Brunner, 2016), see Figure 5.8A. However, model sensitivity was examined for grid spacings of 60 m, 90 m, and 120 m. The working domain of the floodplain was defined using the HEC-RAS mapper as a closed polygon to develop the 2D model grids for the specified spacing. Breaklines along the major rivers were included to adjust the cell face, align the mesh, and to have smooth mesh transition from the rivers to the floodplain. Moreover, the model was simulated for the 2D diffusion wave and 2D Saint Venant equations for comparison of results even though, Brunner et al. (2015) and Quirogaa et al. (2016) indicated that the 2D diffusive wave equations are computationally more efficient than the full 2D Saint Venant equations with similar results and greater stability. The model was also simulated for 30 minutes, 1 hour, and 2 hours computational time steps to determine its stability and sensitivity related to inundation extent. The final model simulation parameters were selected based on stability and the required time for model simulation.

The HEC-RAS 2D model with selected modeling parameters was simulated for the pre-dam 2010 discharge time-series of Ribb and Gumara Rivers, generated by the HEC-HMS model, and the Lake Tana water level as the upstream and downstream boundary conditions, respectively. The model was calibrated against the flood inundation surface area determined from 01 August 2010 Landsat imagery reflectance values. Manning's roughness coefficient ($\eta$) (Table 5.3) was selected as the calibration parameter and the hydrodynamic model was simulated by changing the average roughness values until the best model prediction was obtained. The performance of the hydrodynamic model to capture the retrieved flood inundation area was analyzed using the measure of fit (Bates and De Roo, 2000) given by Equation 5.14.

$$ F = \frac{A_{HEC} \cap A_{GEE}}{A_{HEC} \cup} * 100 \tag{5.14} $$

where $A_{HEC}$ and $A_{GEE}$ are the wet areas predicted by HEC-RAS and GEE, respectively, $\cap$ and $\cup$ are the mathematical operators for HEC-RAS and GEE inundation area intersection and union, respectively, and computed using the ArcGIS software. The value of F ranges from 0 to 100% representing no and perfect matches, respectively.

Table 5.3. Manning's roughness coefficient for different land uses.

| Land use | Area based on land use (km$^2$) | Average Manning's $\eta$ (-) | Source |
|---|---|---|---|
| Cultivated land | 297.28 | 0.04 | Chow (1959) |
| Farm village | 24.54 | 0.1 | Rendon et al. (2012) |
| Forest land | 3.64 | 0.15 | Chow (1959) |
| Grassland | 1.01 | 0.03 | Chow (1959) |
| Shrub and bushland | 1.11 | 0.1 | Chow (1959) |
| Water body | 18.33 | 0.042 | Rendon et al. (2012) |
| Wetland | 37.09 | 0.15 | Rendon et al. (2012) |

The HEC-RAS model does not explicitly simulate infiltration and evaporation of flooded areas (Brunner, 2016). This affects the simulated flooding characteristics of the wetlands and therefore the dam impact on their ecological conditions as flood water cannot recede from such depressions. For this, the rate of seepage and evaporation across the Shesher and Welala wetlands was determined based on literature and generated as a source/sink discharge time-series. The rate of evaporation was equated to the monthly average water surface evaporation of Lake Tana from SMEC (2008a), while the rate of infiltration was taken as 0.5 mm/hr based on soil characteristics described by Berhanu et al. (2013). The computed time-series were used as an internal boundary condition for model simulation. Note, the discharge contribution of the central ungauged watershed to the Fogera Plain was not included for the hydrodynamic model simulation, as it does not have a defined channel outlet.

## Model application

The calibrated hydrodynamic model was simulated from 2010 to 2019 for the pre- and post-dam scenarios to determine the inundation extent, water depth and flood duration at the floodplain and wetlands. For each year, the model simulation period ranged from 01 April to 30 November to study the inundation dynamics. The pre- and post-dam flooding characteristics of the Fogera Plain were determined using ArcGIS 10.5.1 on 01 August of each year for comparison and, to determine the effectiveness of the dam to reduce the flooding parameters. The hydrodynamic model was also used to simulate only for the Ribb and the Gumara River discharge time-series to understand the connectivity of the wetlands to these rivers.

## 5.4.5 Effect of the Ribb Dam on the Fogera Plain Wetlands

The pre-dam ecological condition of the Fogera Plain, especially the status of the wetlands, was identified based on the literature review such as Negash et al. (2011), Wondie (2018), Mohammed and Mengist (2019), and others. These include the assessment of economic and ecological benefits of the wetlands, the lateral connectivity of the wetlands with the major rivers and Lake Tana, the dominant fish types, and the type and characteristics of the birds as described in Section 2.6.3. The pre- and post-dam hydrodynamic model simulation results on 01 August of each year were compared for the spatial and temporal flooding extent, water depth, and duration on the floodplain and the wetlands using ArcGIS. The fish migration path (i.e. the lateral connectivity of the wetlands to the major rivers and Lake Tana) was assessed for the threshold water depth equal to or greater than ($\geq$) 0.5 m as suggested by Limbu (2020) and Abdel-Hay et al. (2020) for pond farming of Clarias gariepinus fish. Specifically, the pre- and post-dam flood duration for the water depth of $\geq$ 0.5 m was developed for the Shesher and Welala wetlands to examine its effect on the spawning and reproduction of the common fish species in Fogera Plain wetlands. Generally, the analysis was used to characterize the ecological effects of the Ribb Dam construction on the aquatic life of the Fogera Plain wetlands.

## 5.5 RESULTS

### 5.5.1 Hydrological Modeling

**Model Calibration and Sensitivity Analysis**

Table 5.4 lists the physical characteristics of the reaches (Rs) and sub-watersheds (SB) of the Ribb and Gumara watersheds, which were produced using HEC-GeoHMS (Figure 5.2). The data were used as input for HEC-HMS modeling.

Table 5.4. Sub-watershed and reach characteristics at the Lower gauging stations for the Ribb and Gumara watersheds.

| Watershed | Sub-watershed/ Reach name | Sub-watershed area (km$^2$) | Reach length (km) | Time of concentration (hr) |
|---|---|---|---|---|
| | SB-01 | 427.42 | | 6.70 |
| | SB-02 | 404.39 | | 5.73 |
| | SB-03 | 218.90 | | 4.01 |
| | SB-04 | 449.87 | | 4.95 |
| | Rs-01 | ------- | 20.25 | ------- |
| Ribb | Rs-02 | ------- | 28.56 | ------- |
| | SB-01 | 547.82 | ------- | 7.73 |
| | SB-02 | 381.56 | ------- | 5.60 |
| | SB-03 | 422.07 | ------- | 7.85 |
| Gumara | Rs-01 | ------- | 29.62 | ------- |

The TP analysis shows that the Ribb watershed surface runoff is mainly dominated by the rainfall measured at Addis Zemen ($A_i$ = 555 km$^2$) and Debre Tabor ($A_i$ = 945 km$^2$) meteorological stations, while the Gumara watershed runoff is dominated by the rainfall measured at the Debre Tabor ($A_i$ = 752 km$^2$) and Wanzaye ($A_i$ = 599 km$^2$) meteorological stations.

Table 5.5. Results of HEC-HMS model calibration for the Ribb (01 June to 30 Sep. 2007) and Gumara (01 June to 30 Sep. 2008) watersheds.

| Watershed | Parameters | Observed | Simulated | Difference | % Diff | NSE | R² |
|---|---|---|---|---|---|---|---|
| Ribb | Volume (Million m³) | 369.1 | 364.3 | -4.8 | -1.3 | | |
| | Peak Flow (m³/s) | 95.7 | 102.6 | 6.9 | 7.2 | | |
| | Hydrograph | | | | | 0.7 | 0.71 |
| | Time of Peak | 24-Aug-07 | 29-Aug-07 | | | | |
| Gumara | Volume (Million m³) | 1047.5 | 971.9 | -75.6 | -7.2 | | |
| | Peak Flow (m³/s) | 288.2 | 272 | -16.2 | -5.6 | | |
| | Hydrograph | | | | | 0.9 | 0.82 |
| | Time of Peak | 5-Aug-08 | 6-Aug-08 | | | | |

HEC-HMS model calibration at the Lower Gauging stations shows a good agreement with NSE values of 0.7 and 0.9 and R² values of 0.71 and 0.82 for the Ribb and Gumara watersheds, respectively. The model captured well the observed peak discharge values, their time of occurrence, the discharge volumes (Table 5.5), and the general temporal trend of the hydrographs for the Ribb and Gumara Rivers (Figure 5.3). Similarly, model validation shows a good agreement between discharge volumes and peak values (Table 5.6). Figure 5.4 shows the simulated and observed hydrographs for the model validation period.

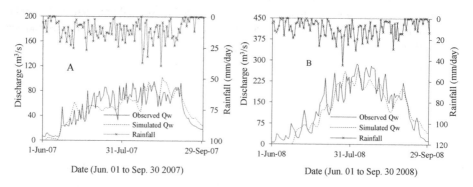

Figure 5.3. Plots of average rainfall and hydrographs of observed and simulated daily river discharge (Qw) for the model calibration period: (A) Ribb and (B) Gumara watersheds.

Table 5.6. Results of HEC-HMS model validation for the Ribb (01 June to 30 Sep. 2008) and Gumara (01 June to 30 Sep. 2009) watersheds.

| Watershed | Parameters | Observed | Simulated | Difference | % Diff | NSE | $R^2$ |
|---|---|---|---|---|---|---|---|
| Ribb | Volume (Million m$^3$) | 363.1 | 412.3 | 49.3 | 13.6 | | |
| | Peak Flow (m$^3$/s) | 118.3 | 112.7 | -5.6 | -4.7 | | |
| | Hydrograph | | | | | 0.6 | 0.69 |
| | Time of Peak | 23-Aug-08 | 8-Sep-08 | | | | |
| Gumara | Volume (Million m$^3$) | 795.2 | 995.7 | 200.6 | 25.2 | | |
| | Peak Flow (m$^3$/s) | 266.9 | 328.1 | 61.2 | 22.9 | | |
| | Hydrograph | | | | | 0.8 | 0.92 |
| | Time of Peak | 5-Aug-09 | 6-Aug-09 | | | | |

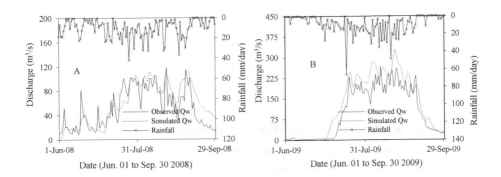

Figure 5.4. Plots of average rainfall and hydrographs of observed and simulated daily river discharge (Qw) for the model validation period: (A) Ribb and (B) Gumara watersheds.

The model simulation shows that the most sensitive parameter affecting the peak and the volume of discharge values for the Ribb watershed is Ca, constant loss, followed by Ra, the storage coefficient. Changing the calibrated Ca value (Table 5.7) by -30% and +30% will decrease and increase the peak discharge by 29.5% and 22.4%, respectively. The average lower and upper bound, and the standard deviation values for the most uncertain parameter for the Ribb sub-watershed (Ca) is 0.0013 mm/hr, 1 mm/hr and 0.2856 mm/hr, respectively, as determined using Monte Carlo analysis of HEC-HMS manager for 1000 trials. However, for the Gumara watershed Ra is the most sensitive parameter followed

91

by Ca. It was observed that changing Ra from the calibrated value by -30% and +30% will decrease and increase the peak discharge by 9% and 6.7%, respectively. The average lower and upper bound, and the standard deviation values for the most uncertain parameter for the Gumara sub-watershed (Ra) are 116.94 hr, 283.23 hr and 48.19 hr, respectively.

Table 5.7. Calibrated watershed and reach parameters for the Ribb and Gumara watersheds

| Watershed and Reach parameters | Unit | Watershed/ Reach name | Calibrated parameter for the: | |
|---|---|---|---|---|
| | | | Ribb watershed | Gumara watershed |
| Initial Loss (Ia) | mm | ALL | 13.359 | 71.83 |
| Constant Rate (Ca) | Mm/hr | ALL | 0.457735 | 0.00500435 |
| Time of Concentration (Tc) | hr | SB-01 | 11.925 | 4.4201 |
| | | SB-02 | 25.141 | 12.709 |
| | | SB-03 | 5.6773 | 150.72 |
| | | SB-04 | 10.516 | |
| Storage Coefficient (Ra) | hr | SB-01 | 402.9 | 164.02 |
| | | SB-02 | 985.02 | 150.72 |
| | | SB-03 | 164.9 | 233.54 |
| | | SB-04 | 175.71 | |

## Model Application for the Pre- and Post- dam Discharge

The hydrological models of the Ribb and the Gumara watersheds with the calibrated and validated watershed and reach parameters were used to generate the time-series discharge values at the Lower Gauging stations between 2010 and 2019. Model simulation resulted in larger peak discharge values and runoff volumes than those measured. As a result, at the location of the gauging station only part of the total flood flow is measured. For example, Figure 5.5 shows the comparison between the measured and the simulated discharge values of the 2010 pre-dam scenario. It is found that the model simulation resulted in larger peak discharge values of 214.0 m$^3$/s and 393.9 m$^3$/s for the Ribb and Gumara Rivers, whereas the measured values were 127.7 m$^3$/s and 279.1 m$^3$/s, respectively. The difference between the measured and the simulated discharge values represents the ungauged volume of the flood that flows to the Fogera Plain from the Ribb and Gumara Rivers upstream of the gauging stations. Figure 5.5 shows that the rising limb of the discharge hydrograph for the Ribb and the Gumara Rivers started in June and reached a maximum around the end of August or the start of September.

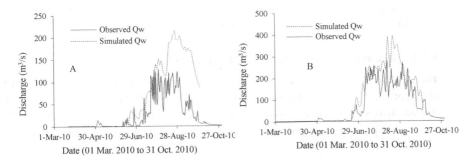

Figure 5.5. Comparison between simulated and observed daily discharge (Qw) at the Lower gauging stations for the Ribb (A) and Gumara (B) watersheds.

The HEC-HMS model was also used to simulate the post-dam discharge of the Ribb River at the Lower Gauging station entering the Fogera Plain for use in HEC-RAS 2D. The Ribb Dam reduced the river discharge magnitude on average by 20% and delayed the outflow by one day. For example, the 24 August 2010 peak inflow discharge value of 130.9 m³/s was reduced by 21% to become 103.9 m³/s and occurred on 25 August 2010.

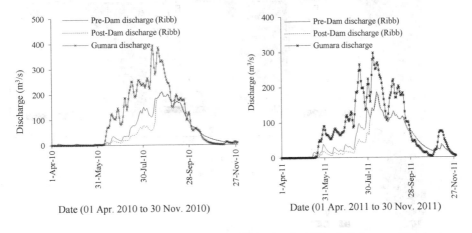

Figure 5.6. Computed pre- and post-dam discharges for the Ribb and Gumara Rivers at the Lower gauging stations near the bridges in (A) 2010 and (B) 2011.

The post-dam discharge of the Ribb River at the Lower Gauging station, which was obtained by combining the spillover discharge with the discharge generated downstream of the dam, is shown in Figure 5.6. It shows first a reduction, as there is no discharge contribution from the upstream watershed of the dam (i.e. dam filling stage) and then rises sharply and approximates the pre-dam discharge as the water level in the reservoir reaches the spillway crest and spillover starts. Both rivers attain their maximum discharge between the end of August and the start of September.

## 5.5.2  Observed Pre-dam Fogera Plain Flooding

The pre-dam flood inundation area was determined based on Landsat image reflectance values of the ground objects using GEE and found to be 195 km$^2$ (Figure 5.7A) on 01 August 2010, which covered about 51% of the Fogera Plain. As stated in Section 5.4.3, the flood map was used as a reference for the calibration of the hydrodynamic model.

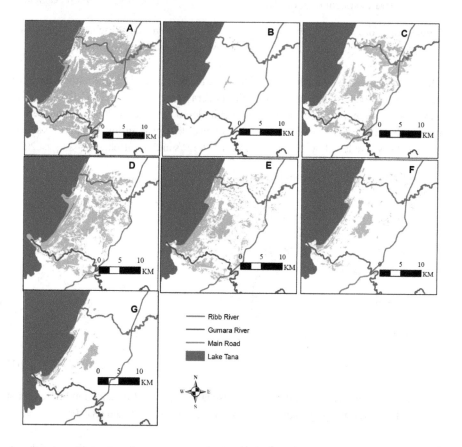

Figure 5.7. Pre-dam flooding of the Fogera Plain determined using GEE for 01 August 2010 (A). The temporal variation of pre-dam flood inundation for the 2011 rainy season: (B) 26 June, (C) 19 July, (D) 20 August, (E) 21 September, (F) 30 September, and (G) 23 October.

The extracted spatial and temporal variations of the open water surface areas for the 2011 rainy season are shown in Figure 5.7(B-G). At the end of the dry period (June), the Welala wetland is not easily visible (Figure 5.7B). An inundation area of 137 km$^2$ is observed in the 3$^{rd}$ week of July (Figure 5.7C), which is more concentrated at the upper, roadside part of the floodplain and along the edge of the rivers. The inundation area has increased to 183 km$^2$ in August (Figure 5.7D) and includes the lower edge of the plain adjacent to Lake Tana. The flooded area is then reduced to 136 km$^2$ one month later (Figure 5.7E). Inundation areas of 59 km$^2$ and 44 km$^2$ are observed in September (Figure 5.7F) and October (Figure 5.7G), respectively, and the flooding is mostly limited to depressions, wetlands, and along the Lake Tana shoreline. The spatial and temporal flooding variation for the 2011 rainy season is presented here to understand the variation throughout the year.

### 5.5.3  HEC-RAS 2D Modeling of the Pre- and Post-dam Flooding

**Model calibration and sensitivity analysis**

The pre-dam hydrodynamic model calibration began by simulating for the average Manning's η based on land use (Table 5.3). It resulted in an inundation area of 162 km$^2$ on 01 August 2010 simulated for 90 m grid spacing and 1 hr. computational time step. Model simulation using +50% and +100% Manning's η resulted in an inundation area of 183 and 188 km$^2$ on 01 August 2010, respectively. This shows that the model is sensitive to the roughness coefficient with an increased area of inundation for increased roughness. The model performance was assessed using the measure of fit (Equation 5.14) and a 46%, 51%, and 52% agreement was obtained between the observed and modeled inundation area for the average, +50% and +100% Manning's η modeling scenarios, respectively (Table 5.8). Manning's η was not increased beyond +100% as the performance of the model does not show much improvement and, the roughness values will be out of the recommended limits, and it is not reduced from the average value as the model performance becomes poorer. The model simulation for the full 2D Saint Venant equations resulted in a similar flood extent as the diffusion-wave shallow water equations indicating that the flooding extent is controlled by the pressure and bottom friction forces. The model results were found less sensitive to the computational time step, while they are sensitive to grid spacing. Hence, further hydrodynamic model simulations were performed for a 100% increased Manning's η, 2D diffusion-wave equation, 90 m grid spacing and, 1 hr. computational time step as it shows a better modeling agreement (52%) and it is more stable with less model simulation time. The final model comprises 49,063 cells with an average face length and cell size of 89 m and 7,973 m$^2$, respectively Figure 5.8 A.

Table 5.8. Comparison of August 1, 2010 flood inundation extent extracted from Landsat imagery and that simulated by HEC-RAS 2D. Note: the inundation area retrieved from Landsat imagery ($A_{GEE}$) is 195 km$^2$.

| HEC-RAS 2D model scenarios for Manning's roughness coefficient ($\eta$) | Simulated area of inundation ($A_{HEC}$) (km$^2$) | $A_{GEE} \cap A_{HEC}$ (km$^2$) | $A_{GEE} \cup A_{HEC}$ (km$^2$) | Measure of Fit, F (%) |
|---|---|---|---|---|
| Average $\eta$ | 162 | 107 | 233 | 46% |
| 1.5*Average $\eta$ | 183 | 125 | 243 | 51% |
| 2.0* Average $\eta$ | 188 | 128 | 245 | 52% |

## Model simulation for the pre- and post-dam flooding

Simulation of a calibrated hydrodynamic model for the 2010 post-dam time-series discharge data yields an inundation area of 165 km$^2$. On average, the Ribb Dam operation resulted in an 11% reduction in flooding extent on 01 August compared to the pre-dam scenario (model comparison from 2010 to 2019). Similarly, the average post-dam inundation area for the water depth $\geq 0.5$ m was reduced by 19.7%, the pre- and post-dam being 79 km$^2$ and 61.4 km$^2$, respectively. In both modeling scenarios, the maximum water depth occurred mainly in the rivers, gullies, wetlands, and depression areas and it was concentrated near the lake shore. Figure 5.8 C shows the simulated post-dam inundation extent for 01 August 2010, which was reduced by 12.2% compared to the pre-dam scenario (Figure 5.8B), while Figure 5.8D shows the overlay of the pre- and post-dam flooding extents. The post-dam model simulation shows ecologically insignificant velocity change from the pre-dam situation (mainly for fish migration), in both cases, it remains less than 0.6 m/s and the maximum flow velocity is observed along the rivers and tributaries. For example, on 01 August 2010, the pre- and post-dam flow velocity of the Ribb River near the Lower Gauging station changes from 0.63 m/s to 0.52 m/s, respectively, while the flow velocity of the Gumara River does not show any changes, as it is not affected by regulation. Moreover, the model results represent well the temporal flood variability, which is important to determine maximum flow parameters such as flow velocity and depth.

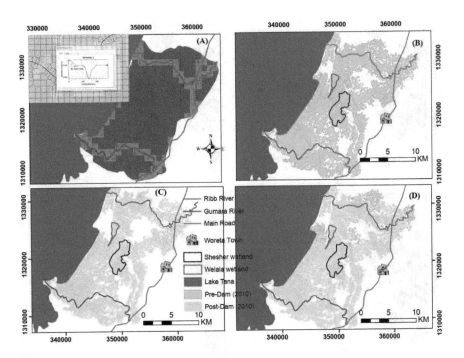

Figure 5.8. (A) Developed grid for HEC-RAS 2D model simulation with detail grids near the Ribb Bridge and profile of river cross-section, (B) Pre- and (C) Post-dam flooding extent of the Fogera Plain determined using HEC-RAS 2D for 01 August 2010, while (D) is the overlay of pre- and post-dam flooding extents.

The pre- and post-dam hydrodynamic model simulation using only the Ribb River discharge data shows that the floodwaters of the river feeds both wetlands (Figure 5.9A), though flooding of the Shesher wetland was reduced in the post-dam scenario (Figure 5.9B). The Gumara River floodwaters also supply both wetlands (Figure 5.9C). The post-dam flooding overlay shows that the central part of the Fogera Plain was supplied by both rivers (Figure 5.9D). Dam operation reduces the pre-dam flooding contribution of the Ribb River from 114.2 $km^2$ to 81.8 $km^2$, while the flooding extent contribution by the Gumara River alone was 119.3 $km^2$ on 01 August 2010.

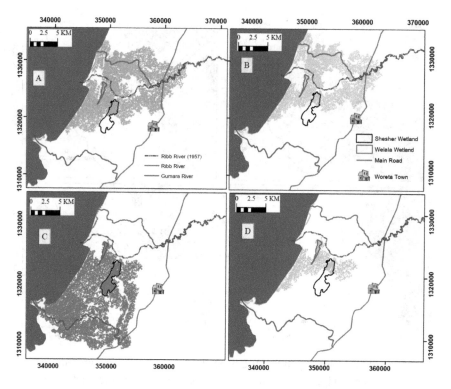

Figure 5.9. (A) Pre- and (B) Post-dam flood inundation by the Ribb River, (C) Flooding by the Gumara River and (D) Post-dam flooding extent shared by the two rivers in 01 August 2010.

The effect on the floodplain of the Ribb River discharge reduction was assessed by comparing the pre- and post-dam discharge hydrographs near the Lower Gauging station with the inflow hydrograph (pre- and post-dam) at the start of the floodplain (Figure 5.10). It shows that the peak discharge near the Lower Gauging station for the pre- and post-dam condition is similar and did not exceed 110 m³/s, which is the estimated bankfull discharge at the location as determined by Mulatu et al. (2018). That means the discharge above the bankfull starts to spill to the floodplain before reaching the Lower Gauging station. The pre- and post-dam discharge hydrograph near the Lower Gauging station was obtained from HEC-RAS model simulation.

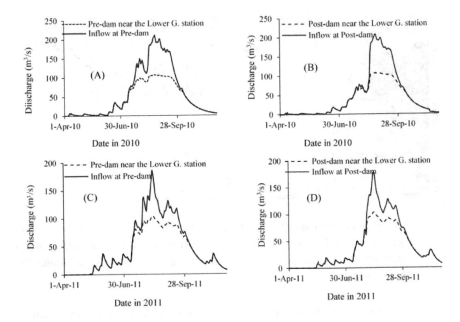

Figure 5.10. Flow hydrograph at the start of the floodplain and near the downstream
gauging station for the pre- and post-dam. (A) and (B) for 2010 and, (C) and (D)
for 2011.

### 5.5.4 Effect of River Regulation on Wetland Dynamics

The pre- and post-dam flood inundation extent for a water depth of $\geq 0.5$ m (Figure 5.11),
developed for 01 August 2010, was used to study the lateral connectivity of the wetlands
to the Ribb and Gumara Rivers and Lake Tana. This helps to determine the possible
pathways of fish to the wetlands. It is observed that the post-dam flooded area with water
depths $\geq 0.5$ m was reduced by 26.8%, the pre- and post-dam flooded areas being 109
km$^2$ and 80 km$^2$, respectively. The inundation maps were not able to resolve defined
channel/pathway connecting the wetlands to the rivers and Lake Tana. Rather, it shows
only flows through the floodplain during flooding as indicated in Figure 5.11A and B for
the pre-and post-dam modeling scenarios, respectively. The Shesher wetland has a
connection with the Gumara River and Lake Tana for the pre- and post-dam scenarios.
However, the Welala wetland may lose its lateral connectivity to the Ribb River for the
post-dam scenario as it receives water only from Lake Tana.

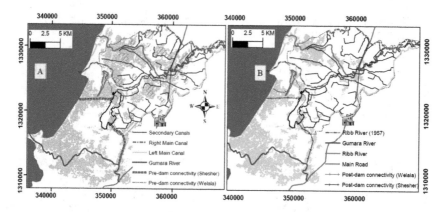

Figure 5.11. Fogera Plain flooding extent for a water depth equal to or greater than 0.5 m
for (A) Pre- and (B) Post-dam scenarios on 01 August 2010. The solid red polygon
delineates the Welala wetland, while the solid black polygon delineates the
Shesher wetland.

The pre- and post-dam hydrodynamic model analyses for a water depth $\geq$ 0.5 m showed
that the flood duration for the Shesher and Welala wetlands extends up to four months as
shown in Figure 5.12. The pre- and post-dam flooding extent covers 1.7% and 0.4% of
the flooded area of the Shesher wetland for a flood duration of greater than or equal to
three months, respectively. 17.4% of the pre-dam and 24.4% of the post-dam flooded
Shesher wetland area has a flooding duration greater than or equal to 60-days. Dam
operation does not affect the Shesher flooding extent (~46% of its flooded area) for a
flood duration of less than 30-days. Hence, the Shesher wetland area that is flooded for
more than 30 days is approximately 54% of the wetland flooded area for both pre- and
post-dam scenarios. The pre- and post-dam flooding extent of the Welala wetland for a
duration of greater than or equal to 90-days covers 3.6% and 5% of the wetland flooded
area, respectively. The pre- and post-dam Welala flooding extent for a duration of greater
than or equal to 60-days is 93% and 99.2% of the wetland flooded area. Note that the
flooded area of the Welala wetland for flood durations less than one month is negligible
for both the pre- and post-dam cases. The Ribb Dam operation increases the flood extent
for the duration of up to 60-days of the Shesher and Welala wetlands by 7% and 6.2%,
respectively, while it shows a negligible effect for flood durations less than 30-days and
exceeding 90-days.

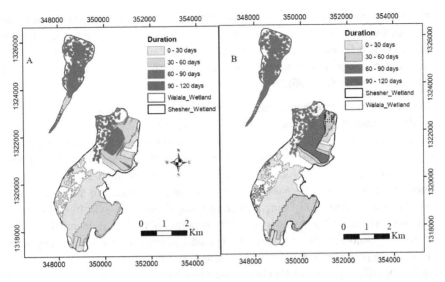

Figure 5.12. (A) Pre- and (B) post-dam flood duration for the Welala and Shesher
wetlands for water depth equal to or greater than 0.5 m for the 2010 rainy season.

## 5.6 DISCUSSION

The HEC-HMS hydrological model calibration and validation showed good agreement
between the measured and simulated discharge time series values as assessed using NSE
and $R^2$ (Section 5.5.1). However, a slight over and under prediction of the low and high
flows were observed (Figure 5.3 and Figure 5.4), which is a common problem for
hydrological models (Zhang and Savenije, 2005). Moreover, the results of hydrological
modeling may have uncertainties related to model input data, model parameters and,
model selection (Sharafati et al., 2020). River discharge computation based on staff gauge
height reading is another source of uncertainty as the rating curves may not be updated
regularly for the alluvial river section, which is affected by the river bank and bed
morphological adjustments. The model analysis confirms that the most uncertain
parameter for the Ribb and Gumara watersheds are the constant rate and the storage
cofficient, respectively.

Obtaining historical flooding information is difficult and uncertain task in data-poor
regions, but valuable for the calibration and validation of hydrodynamic models. Due to
the advances in remote sensing and computer technology, the historical spatial and
temporal variations of inundation extent can be obtained by analyzing multiband satellite
images using a cloud computing platform to obtain grid-based flooding information. The
derived inundation map was used to calibrate the hydrodynamic model and study the pre-
and post-dam flooding dynamics of the Fogera Plain. The method was also employed to

monitor the spatial and temporal inundation of the floodplain and the wetlands. However, the detection of clear water inundation may be affected by cloud cover, topographically-induced shadow effects (Donchyts et al., 2016; Teng et al., 2017), and the resolution of the satellite image where mixed land uses in coarse image resolutions affect the reflectance values (Xu, 2006).

The developed hydrodynamic model was found more sensitive to the roughness coefficient than the grid spacing. The model simulation for the upper limit of the Manning's η values (+100% of the average) resulted in a greater inundation extent and a higher water depth compared to the modeling results of the average Manning's η values, which was also noted by Brunner (2016) and Pinos and Timbe (2019). However, it is insensitive to the computational time steps of a similar mesh size. In this study, model simulation for a 90 m grid spacing, 1 hr computational time step, and +100% Manning's η showed an acceptable result with a 52% measure of fit for the inundation extents obtained by remote sensing. However, better model performance was obtained by others, for example (i) Pinos and Timbe (2019) found an agreement ranging from 40.7% to 88.2% for different zones of flooding of the Santa Barbara River Basin in the southern part of Ecuador using a flood inundation map produced by a previous HEC-RAS 1D model simulation as a base case; and (ii) Quirogaa et al. (2016) found 72% and 73% inundation agreement between the flooding extent obtained from the moderate resolution imaging spectroradiometer (MODIS) satellite and the HEC-RAS 2D model with varying Manning's η values for Llanos de Moxos floodplain in Bolivian Amazonia. The relatively low model agreement with the inundation extent retrieved using Landsat satellite images may be related to (i) the quality of hydrodynamic model input data including the rivers' discharge time-series data, the roughness coefficient, the grid spacing, and the resolution of the underlying topography (DEM) (Oubennaceur et al., 2018; Vojtek et al., 2019); (ii) the accuracy of the method to determine the inundation extent using remote sensing (Donchyts et al., 2016; Teng et al., 2017); (iii) the discharge contribution from the ungauged watersheds by overland flow; and (iv) the direct precipitation on the Fogera Plain. The inability of HEC-RAS 2D version 5.0.7 to consider the effects of evaporation and infiltration was resolved by developing, a time series (negative) discharge that is the sum of evaporation and infiltration for each wetland and imposing it as an internal boundary condition at the lowest-elevation grid cell of the respective wetland.

The Ribb Dam operation was found to attenuate the outflow discharge on average by 20%. A study by Mulatu et al. (2020) using average monthly discharge data showed a similar result in which the dam operation resulted in a reduced wet season peak discharge value by 34%. The hydrologic analysis conducted by WWDSE and TAHAL (2007) for the spillway design also indicated a reduction of maximum outflow discharge by 38%. However, the pre- and post-dam hydrodynamic model simulation showed that the Ribb Dam has a negligible effect on Fogera Plain flooding as it regulates only 23.8% of the total watershed area that contributes flow to the plain. The recent 2019 flooding (Figure

5.13), which occurred in the middle of August, confirms the ineffectiveness of the dam to reduce flooding. According to regional (Amhara Mass Media Agency) and federal (Ethiopian Broadcast Centre) media sources, the flooding was caused by heavy rain in the Ribb watershed, and resulted in three deaths and the inundation of ~4,700. During the flooding event, the watershed above the dam was not contributing discharge as it was in the filling stage. This indicates that dam construction at the upper reach of the river system to regulate a small portion of watershed area may have limited influence to attenuate the discharge at the lower reach due to additional discharge from the downstream ungauged watersheds.

Figure 5.13. Flooding on the Fogera Plain in the main rainy season of 2019 (A) downstream view on the main road from Woreta to Addis Zemen near the Ribb River and (B) inundated villages near the main road around the Ribb River.

The 2010 pre-dam model result showed water depths exceeding 0.5 m covers 55.5% of the inundated area, while this area is reduced to 46.2% for the post-dam scenario. Similarly, the 2010 pre-dam inundated area for water depths exceeding 1 m covers 18.1% and it is reduced to 14.2% in the post-dam case. The lag time presented between the maximum Lake Tana water level and the peak discharge of the rivers also helped to reduce the flooding extent. The occurrence of maximum Lake Tana water level in late September resulted in retarded drainage of the plain and an increased flood duration of the wetlands (Figure 5.7F). The post-dam model simulation also showed an insignificant flow velocity reduction compared with the pre-dam case, which is less than 0.6 m/s in both scenarios.

The Ribb Dam operation will alter the normal river discharge causing reduced and increased wet and dry season values, respectively. The construction of the irrigation system with canals, river bank protection, and water regulatory works may create a barrier that restricts the movement of water and fishes (Figure 5.11). This may disrupt habitat dynamics, lower the ecosystem productivity, and affect the livelihood of dependent dwellers (Richter et al., 2003). Though the post-dam model simulation shows negligible flood extent reduction of the Fogera Plain, the release of excess water from farmlands in the dry season may shift the seasonal Shesher and Welala wetlands to permanent as they are planned to be used as a detention/collection basin (Figure 5.11). This could increase fisheries and attract birds and other aquatic communities if the water quality is kept in good condition. Fishing may also be practiced in the Ribb reservoir based on observations from the nearby Koga reservoir.

## 5.7 CONCLUSIONS

This study assessed the hydrological impact of the Ribb Dam on the Fogera Plain particularly on the flooding dynamics, using the HEC-RAS 2D hydrodynamic model. The results of the model were then used to discuss the alteration of ecologically relevant flow parameters which are vital to study ecological consequences for the Fogera Plain wetlands. The HEC-HMS hydrologic model was used to generate the time-series discharge for the Ribb and Gumara Rivers that flow into the Fogera Plain, including the influence of the Ribb Dam, as the existing measuring stations are unable to gauge the overbank flow portion. Hydrological model calibration for the Ribb and Gumara watersheds showed an NSE value of 0.7 and 0.9, respectively, while 0.6 and 0.8 for validation.

The application of HEC-HMS for the determination of pre- and post-dam discharge data for the watershed was well understood in the literature. Moreover, the Ribb Dam also controls a small portion of the watershed that drains to the Fogera Plain and the flooding attenuation of the plain was minimal as determined. However, it is important for the region as there is no study done to (1) determine the degree the dam will affect the flooding extent and (2) associate the dam-induced hydrological changes to the ecological consequences that may come due to impoundment at the headwater of the Ribb River. The results of the hydrologic model simulation showed that dam construction at the upper reach of the Ribb River will attenuate the peak discharges, on average by 20%. At the lower gauging station, the natural discharge hydrograph is almost recovered due to flow contributions from the downstream watershed. Moreover, the model can be used to study future watershed development scenarios, including land use and land cover, and climate change.

In this study, the cloud computing platform of the GEE is applied to determine the historical inundation maps from the Landsat image reflectance values as historical flood inundation maps are lacking for the region. The mapped flooded areas were then used for the calibration of the hydrodynamic model. The method was found less costly and less time-consuming to map flood inundation mapping with an acceptable level of confidence for data-poor areas.

The results of the hydrodynamic model simulation indicated that the dam does not significantly reduce flooding of the Fogera Plain. The flooding due to Lake Tana backwater dominates in the lake shore area and propagates up the main rivers. The proposed dam operation negligibly affects the lateral connectivity of the Shesher and Welala wetlands (indicative path for fish migration) to the major rivers and Lake Tana. The probable increase in post-dam dry-time flow to the wetlands from excess irrigation water could further sustain the availability of fish and bird species.

The accuracy of the floodplain-terrain representation in the HEC-RAS 2D hydrodynamic model and the boundary conditions may have a significant impact on the results. For this study, a freely available 30 m resolution DEM was sufficient to determine the inundation extent. The post-dam hydrodynamic model simulation showed that the Ribb Dam will reduce the Fogera Plain flooding extent of a water depth exceeding 0.5 m by 26.8%. The results do not show ecologically significant changes in flooding extent and duration of the Shesher and Welala wetlands, which are vital for the survival of the original habitat. However, the developed hydrodynamic model can be used to study different dam operation scenarios to evaluate the change of flooding extent, depth, duration, arrival time, etc. easily at a specific location.

# 6

# CONCLUSIONS AND RECOMMENDATIONS

The aim of this study is to anticipate the hydrological, morphological and ecological effects of dam construction on downstream river reaches, focusing on the Ribb River. The main results are presented in the previous chapters, while this chapter deals with the conclusions based on overall integration of the results and recommendations for further study.

## 6.1 CONCLUSIONS

This study provided answers to the general questions developed in Section 1.3 as follows:

---

**What were the morphological trends of the Ribb River before damming?**

---

The collection of primary and secondary data, including the river channel geometry, time-series river discharge, water levels, aerial photographs, SPOT, and Google Earth satellite images was essential for the analysis of the morphological trends of the river system before damming. The analysis of data and the application of the Equilibrium Theory by Jansen et al (1979) show that the Ribb River system experienced channel avulsion and cut-offs, bank migration, and bed level rise (Chapter 3). The reach-averaged sediment transport capacity becomes less becomes less going downstream. This increased the height of the riverbed and in 2008 avulsion of the channel downstream of the Ribb bridge. Sand mining reduces the bed slope. The dry season water extraction for irrigation has had negligible effects on the river slope. Finally, the Lake Tana level regulation for power production has only a trivial impact on the closure of the old river channel, as analyzed using the De Vries (1975) method.

---

**How will the Ribb Dam operation affect the river discharge regime and the Fogera Plain flooding extent?**

---

The completion of the Ribb Dam and the beginning of operation may affect and change the current morphodynamic trends of the Ribb River in a less active manner due to reduced peak discharge values. However, the release of less sediment-laden flow from the dam will cause the river bed degradation and, river bank erosion in the downstream river reaches of the dam. The long-term hydrological impacts by the Ribb Dam on the downstream river reaches were investigated considering three dam operation scenarios based of the size of the irrigated area (Chapter 4). The results show that the dam will have a distinct hydrological effect, including the change of the daily maximum and minimum values with a time shift of the low and high discharge values, and a decrease in yearly discharged volume below the weir. The HEC-HMS hydrologic model simulation also showed that Ribb dam operation will attenuate the natural peak discharge of the river at the dam site, on average by 20%. However, the natural discharge hydrograph is less affected at the lower gauging station due to flow contributions from the downstream watersheds (Chapter 5).

The effects of Ribb Dam operation on the Fogera Plain flooding extent were studied using HEC-RAS 2D hydrodynamic models (Chapter 5). The flooding extent generated by the spectral reflectance difference values of the ground object of the Landsat satellite images retrieved using the Google Earth Engine was used to calibrate the 2D hydrodynamic model for the pre-dam scenario. The model simulation shows that the Ribb Dam operation would be found to reduce the extension of the flooded area of the Fogera Plain by 11%, on average. The comparison of pre- and post-dam model simulations result allows to assess the change in the water depth, and flow velocity on the Fogera Plain. Due to reduced post-dam river discharge, the model simulations show a slight reduction of water depth and flow velocity. This indicates that the Ribb Dam operations will have only a modest effect on the overall flood extent, depth and, velocity of the Fogera Plain. Hence, there may be a need to construct levees at the selected locations of the river banks for flood protection.

> **What are the major factors related to flooding that influence the Fogera Plain ecology and to what extent will the operation of the Ribb Dam affect them?**

Even the modest alterations of ecologically relevant discharge values (in terms of water availability, flooding duration, and flow velocity) of the Fogera Plain wetlands due to Ribb Dam operation may still have an affect on the natural biota. The ecological consequences were analyzed based on the results of the hydrodynamic models comparing the pre- and post-dam conditions (Chapter 5). These were assessed for the threshold water depth equal to or greater than ($\geq$) 0.5 m as suggested by Limbu (2020) and Abdel-Hay et al. (2020) for the Walala and Shesher wetlands; assumed to be a sufficient depth for the fish migration. The analysis shows that the Ribb Dam will reduce the Fogera Plain flooding extent for a water depth exceeding 0.5 m by 26.8%. The pre- and post- dam flood duration on the Fogera Plain wetlands for a water depth of $\geq$ 0.5 m extends up to four months. The dam operation increases the flooded area of the Shesher and Welala wetlands for the duration of 30-60 days, while it shows a negligible effect for flood durations less than 30 days and exceeding 90 days. Moreover, the hydrodynamic model simulation shows that the lateral connection of the wetlands to the major water sources for a threshold water depth of $\geq$ 0.5 m is less affected. The Shesher wetland has a connection with the Gumara River and Lake Tana for the pre- and post-dam scenarios. However, the Welala wetland may lose its lateral connectivity to the Ribb River for the post-dam scenario as it receives water only from Lake Tana.

The construction of irrigation structures on the floodplain will create barriers to the free movement of water and the migration of fish to the wetlands. The usage of pesticides and fertilizers will deteriorate the wetlands' water quality as the irrigation excess water will be released to it. The release of irrigation excess water to the wetlands may increase the water availability of the wetlands in the dry period and create some favorable conditions, if the water quality is maintained well. The dam operation that allocates the environmental flow may also improve the water availability to the downstream reaches during the dry season as the irrigation water requirements will be met with the dam stored water. Currently, irrigators abstract the river flows and it becomes dry in the dry periods.

> **How will the dam operation affect the morphological condition of the downstream river system on the long term?**

The morphological changes downstream of the dam were first assessed using an analytical method based on the Equilibrium Theory of Jansen et al. (1979). The method was applied to the sand- and gravel-bed reaches to describe the new idealized reach-scale morphodynamic equilibrium characteristics with the idea to compare dam operation scenarios in feasibility studies (Chapter 4). Moreover, a 1D morphodynamic model (SOBEK-RE) was run for 1,500 years for comparison of results and, to estimate the required time to reach the new reach-scale equilibrium, assuming the simulation period is long enough to obtain it. Both methods showed that the dam operation scenario that releases the maximum peak discharge values would result in maximum downstream river bed degradation near the dam and, the alteration would propagate downward at a higher rate compared to the other scenarios. The 1D model simulation also confirmed that the sand-bed reach controls the time required to achieve the theoretical new equilibrium, as it showed a higher rate of river bed degradation than the gravel-bed reaches after a long period of dam operation (in this case 1,500 years). However, a large portion of the total bed degradation at the dam base and the weir that requires remedial measures occurs in the first 100 years.

Generally, this comprehensive study showed that the Ribb River has undergone different morphological processes before dam construction. The dam construction to store the wet-season flood, its operation to release a discharge based on the irrigation water demand through the main river channel until the diversion weir site, the release of environmental flow and the diversion of discharge to the command areas for irrigation will create new conditions in the downstream river system. For these, hydrological, hydrodynamic, and morphological models were applied successfully to determine the dam impacts on the downstream river discharge, flooding extent, and morphological conditions, respectively. The application of simplified empirical Equilibrium Theory allows selecting the most impactful dam operation scenarios with less data and time. The developed hydrological, hydrodynamic and, morphological models help to study the impacts of other dam operation scenarios, climate, and land use, and land cover changes. It is also found that the application of Google Earth Engine is vital to determine flood inundation maps for data-scarce regions.

The application of hydrologic and hydrodynamic models shows that the Ribb Dam operation may bring slight changes to the flooding extent, water depth, and flow velocity of the Fogera Plain and its wetlands. This helps to keep the original ecosystem. However, the continued flooding may affect infrastructures and cause fatalities. This indicates that achieving a balance between flood protection and ecological conservation is challenging in a system like the Fogera Plain due to the close and intertwined distribution of people

110

and wetland systems. Hence, an integrated approach that will address the flooding consequences on the ecosystem and the people, the new irrigation perimeter and changes in river morphology, and the property shall be required.

## 6.2 RECOMMENDATIONS

The hydro-geomorphological adaptation of the Ribb River system downstream of the dam was analyzed using primary and secondary data. This study improved the understanding of the hydrological alteration of the river system to the ecology of the Fogera Plain wetlands. However, the following recommendations are developed for further research works on the dam-induced downstream effects.

1. In the absence of discharge measuring stations at the points of interest (in this case, at the end of Upper-I reach, the weir site, and the flood starting point), the daily time-series discharge was generated based on watershed proportionality and using a calibrated hydrological model assuming the watershed properties like the LULC and rainfall pattern does not influence the watershed yields. However, this assumption may give uncertain results as the rainfall pattern and watershed properties vary spatially. Therefore, temporary or permanent hydrological stations should be implemented at the end of Upper-I reach, the weir site, and the flood starting point.

2. The developed hydrological, hydrodynamic, and morphodynamic models can be used to study other watershed management, climate change, and dam operation scenarios. However, as discussed in various parts of the thesis, they are developed with certain assumptions due to the shortage of data. Hence, further research should be done by incorporating new field data and for other hydrological, hydrodynamic, and morphodynamic parameters.

3. The existing hydrological stations in the Lake Tana Sub-basin measures only discharge data. In some instances, suspended sediment samples were taken during the rainy season to develop a rating curve to estimate the yearly budget. Moreover, there is no hydrological station that measures the bed load of the river. However, for this study, the pre- and post-dam sediment transport capacity of the river reaches was done by selecting the appropriate sediment transport capacity equation from literature, which may yield uncertain results. Hence, bed-load sampling should be done at least in the rainy season for better understanding and, to validate the selected empirical sediment transport equations.

4. In developing the one-dimensional morphodynamic model, single grain size and different sediment transport equations were used for each reach. The model was developed considering the bed level changes are dependent on the sediment balance equation (Exner, 1925 principle). The application of this equation creates a sudden transition of gravel grains to sand when they pass the transition point of the gravel to sand reach. However, in the actual case, it is very hard to

find a river in which the bed is occupied by a single grain. Hence, future research should be done considering the river system as a gravel-sand mixture. Furthermore, future works should be done to validate the applied analytical and 1D morphodynamic methods if they allow for detecting the least impacting scenario, considering that data showing the effects of long-term dam operations on the downstream river reaches are lacking.

5. The study indicated that the Ribb Dam operation may bring a slight impact on the ecological significant flow parameters that are important to keep the original aquatic life of the Fogera Plain wetlands. However, there should be a study to determine the water quality of the wetlands after the operation of the irrigation scheme and to propose remedial measures as wetlands are planned to use as a detention basin.

6. The Lake Tana Basin is identified as one of the growth corridors for its significant natural resources (land and water), livestock, and fishery with developed urban centers, road connectivity, and rich cultural heritage. These natural resources are finite and the development plans should be within the limit. Therefore, a deep understanding of hydrological variables and processes, and the spatio-temporal variability of rainfall and water demands are important for the sustainability of water resources developments. The major limitation of past studies is the predominant scarcity of quality data. An increase in the network of hydro-metrological gauging stations at different key locations shall be given priority. Existing gauging stations shall also be maintained and their standards improved, even though some works are already started by the Tana Sub-basin Authority on the major rivers. To address the existing data gaps, the use of remote sensing tools can also be used in line with field-measured data. Future studies should include basin-wide analyses of developed, ongoing, and planned water resources development scenarios. This helps for a better understanding of their effects on the hydrological and morphological conditions and ecological consequences.

# REFERENCES

Abate, M., 2016. *Human impact on Hydro-geomorphology of Gumara River upper Blue Nile basin, Ethiopia.* (PhD), Bahir Dar Institute of Technology, Bahir Dar University, Bahir Dar.

Abate, M., Nyssen, J., Steenhuis, T. S., Moges, M. M., Tilahun, S. A., Enku, T., & Adgo, E., 2015. Morphological changes of Gumara River channel over 50 years, Upper Blue Nile Basin, Ethiopia. *Journal of Hydrology, 525,* 152-164. doi:10.1016/j.jhydrol.2015.03.044.

Abdel-Hay, A. H. M., Emam, W., Omar, A. A., Eltras, W. F., & Mohamed, R. A., 2020. The effects of rearing water depths and feed types on the growth performance of African catfish (Clarias gariepinus). *Aquaculture Research, 51*(2), 616-622. doi:org/10.1111/are.14409.

Abebe, W. B., G/Michael, T., Leggesse, E. S., Beyene, B. S., & Nigate, F., 2017. Climate of Lake Tana Basin. In K. Stave, G. Goshu, & S. Aynalem (Eds.), *Social and Ecological System Dynamics: Characteristics, Trends, and Integration in the Lake Tana Basin, Ethiopia*: Springer.

Abebe, W. B., & Minale, A. S., 2017. Land use and watershed management practices in Lake Tana Basin. In K. Stave, G. Goshu, & S. Aynalem (Eds.), *Social and Ecological System Dynamics: Characteristics, Trends, and Integration in the Lake Tana Basin, Ethiopia* (pp. 479-521): Springer.

Abebe, W. B., Tilahun, S. A., Moges, M. M., Wondie, A., Derseh, M. G., Nigatu, T. A., Mhiret, D. A., Steenhuis, T. S., Camp, M. V., Walraevens, K., & McClain, M. E., 2020. Hydrological Foundation as a Basis for a Holistic Environmental Flow Assessment of Tropical Highland Rivers in Ethiopia. *Water, 12*(2), 547. doi:10.3390/w12020547.

Abera, M., 2017. Agriculture in the Lake Tana Sub-basin of Ethiopia. In K. Stave, G. Goshu, & S. Aynalem (Eds.), *Social and Ecological System Dynamics: Characteristics, Trends, and Integration in the Lake Tana Basin, Ethiopia* (pp. 375-397). Cham: Springer International Publishing.

Ahn, J. M., Jung, K. Y., & Shin, D., 2017. Effects of coordinated operation of weirs and reservoirs on the water quality of the Geum River. *Water, 9*(6), 423.

Ali, Y. S. A., 2014. *The impact of soil erosion in the upper Blue Nile on downstream reservoir sedimentation.* (PhD), IHE Delft, Institute for Water Education, Delft, The Netherlands.

Annys, S., Adgo, E., Ghebreyohannes, T., Van Passel, S., Dessein, J., & Nyssen, J., 2019. Impacts of the hydropower-controlled Tana-Beles interbasin water transfer on downstream rural livelihoods (northwest Ethiopia). *Journal of Hydrology, 569,* 436-448.

Anteneh, W., Dejen, E., & Getahun, A., 2012. Shesher and Welala floodplain wetlands (Lake Tana, Ethiopia): are they important breeding habitats for Clarias gariepinus and the migratory Labeobarbus fish species? *The Scientific World Journal.*

Awulachew, S. B., McCartney, M., Steenhuis, T. S., & Ahmed, A. A., 2009. *A review of hydrology, sediment and water resource use in the Blue Nile Basin* (Vol. 131): IWMI.

Awulachew, S. B., Yilma, A. D., Loulseged, M., Loiskandl, W., Ayana, M., & Alamirew, T., 2007. *Water resources and irrigation development in Ethiopia* (Vol. 123): IWMI.

Aynalem, S., 2017. Birds of Lake Tana Sub-basin. In K. Stave, G. Goshu, & S. Aynalem (Eds.), *Social and Ecological System Dynamics: Characteristics, Trends, and Integration in the Lake Tana Basin, Ethiopia* (pp. 179-205). Switzerland: Springer International Publishing.

Basson, G., 2004. *Hydropower dams and fluvial morphological impacts–An African perspective.* Paper presented at the United Nations Symposium on Hydropower and Sustainable Development, 27–29 October, Beijing, China.

Batalla, R. J., Gomez, C. M., & Kondolf, G. M., 2004. Reservoir-induced hydrological changes in the Ebro River basin (NE Spain). *Journal of Hydrology, 290*(1-2), 117-136. doi:10.1016/j.jhydrol.2003.12.002.

Bates, P., Trigg, M., Neal, J., & Dabrowa, A., 2013. LISFLOOD-FP. In *User manual, University Road, Bristol, BS8 1SS, UK: School of Geographical Sciences, University of Bristol.*

Bates, P. D., & De Roo, A. P. J., 2000. A simple raster-based model for flood inundation simulation. *Journal of Hydrology, 236*(1), 54-77. doi:10.1016/S0022-1694(00)00278-X.

Baxter, R., 1977. Environmental effects of dams and impoundments. *Annual review of ecology and systematics*, 255-283.

BCEOM. 1998. *Abay River Basin Integrated Development Master Plan – Phase 2 –Land Resources Development – Reconnaissance Soils Survey,.* Addis Ababa.

Belete, M. A., 2013. *Modeling and analysis of Lake Tana Sub-Basin water resources systems, Ethiopia.* (PhD), University of Rostock, Germany.

Benda, L., Andras, K., Miller, D., & Bigelow, P., 2004. Confluence effects in rivers: Interactions of basin scale, network geometry, and disturbance regimes. *Water Resources Research, 40*(5), 15. doi:10.1029/2003wr002583.

Berends, K. D., de Jongste, A., van der Mheen, M., & Ottevanger, W., 2015. *Implementation and verification of morphology in SOBEK 3.* Paper presented at the NCR-Days 2015, Nijmegen, Netherlands.

Berhanu, B., Melesse, A. M., & Seleshi, Y., 2013. GIS-based hydrological zones and soil geo-database of Ethiopia. *Catena, 104*, 21-31.

Berz, G., 2000. *Flood disasters: lessons from the past—worries for the future.* Paper presented at the Proceedings of the Institution of Civil Engineers-Water and Maritime Engineering.

Bewket, W., & Sterk, G., 2005. Dynamics in land cover and its effect on stream flow in the Chemoga watershed, Blue Nile basin, Ethiopia. *Hydrological Processes: An International Journal, 19*(2), 445-458.

Beyene, A. A., 2018. *Water balance, extent and efficiency of irrigation in the Lake Tana basin, Ethiopia.* (PhD), Ghent University, Belgium.

Biemans, H., Haddeland, I., Kabat, P., Ludwig, F., Hutjes, R., Heinke, J., Von Bloh, W., & Gerten, D., 2011. Impact of reservoirs on river discharge and irrigation water supply during the 20th century. *Water Resources Research, 47*(3).

Blom, A., Arkesteijn, L., Chavarrías, V., & Viparelli, E., 2017. The equilibrium alluvial river under variable flow and its channel-forming discharge. *Geophysical Research: Earth Surface, 122*(10 ), 1924-1948 doi:10.1002/2017jf004213.

Bolla Pittaluga, M., Luchi, R., & Seminara, G., 2014. On the equilibrium profile of river beds. *Geophysical Research: Earth Surface, 119*(2), 317-332.

Brandt, S. A., 2000. Classification of geomorphological effects downstream of dams. *Catena, 40*(4), 375-401. doi:10.1016/S0341-8162(00)00093-X.

Bravard, J.-P., Kondolf, G. M., & Piégay, H., 1999. Environmental and societal effects of river incision and remedial strategies. *In Incised River Channels, Simon A, Darby SE (eds). John Wiley and Sons: Chichester;, 303–341.*

BRLi, & MCE. 2010. *Pump, drainage schemes at Megech: Environmental and Social Impact Assessment of the Ribb Irrigation and Drainage Project.*

Brune, G. M., 1953. Trap efficiency of reservoirs. *Eos, Transactions American Geophysical Union, 34*(3), 407-418. doi:10.1029/TR034i003p00407.

Brunner, G. W., 2016. *HEC-RAS River Analysis System: 2D Modeling User's Manual. Version 5.0.* US Army Corps of Engineers—Hydrologic Engineering Center.

Brunner, G. W., Piper, S. S., Jensen, M. R., & Chacon, B., 2015. *Combined 1D and 2D hydraulic modeling within HEC-RAS.* Paper presented at the World Environmental and Water Resources Congress.

Childs, M., 2010. Literature survey: The impact of dams on river channel geomorphology. *Departamento de Geografía. Universidad de Hull, Reino Unido.*

Chorowicz, J., Collet, B., Bonavia, F., Mohr, P., Parrot, J., & Korme, T., 1998. The Tana basin, Ethiopia: intra-plateau uplift, rifting and subsidence. *Tectonophysics, 295*(3-4), 351-367.

Chow, V. T., 1959. *Open-channel hydraulics*: McGraw-Hill civil engineering series.

Church, M., 1995. Geomorphic Response to River Flow Regulation: Case Studies and Time-scales. *Regulated Rivers-Research and Management, 11*(1), 3-22. doi:10.1002/rrr.3450110103.

Conway, D., 1997. A water balance model of the Upper Blue Nile in Ethiopia. *Hydrological Sciences Journal, 42*(2), 265-286.

Coulthard, T. J., Neal, J. C., Bates, P. D., Ramirez, J., de Almeida, G. A., & Hancock, G. R., 2013. Integrating the LISFLOOD-FP 2D hydrodynamic model with the CAESAR model: implications for modelling landscape evolution. *Earth Surface Processes and Landforms, 38*(15), 1897-1906.

Crosato, A., 2008. *Analysis and modelling of river meandering.* (PhD thesis), TU Delft, Delft University of Technology, Nieuwe Hemweg 6b, 1013 BG Amsterdam, The Netherlands.

Curtis, K. E., Renshaw, C. E., Magilligan, F. J., & Dade, W. B., 2010. Temporal and spatial scales of geomorphic adjustments to reduced competency following flow regulation in bedload-dominated systems. *Geomorphology, 118*(1), 105-117. doi:10.1016/j.geomorph.2009.12.012.

Damtie, B., Boersma, E., & Stave, K., 2017. Introduction: Regional Challenges and Policy Questions. In K. Stave, G. Goshu, & S. Aynalem (Eds.), *Social and Ecological System Dynamics: Characteristics, Trends, and Integration in the Lake Tana Basin, Ethiopia* (pp. 3-8): Springer.

De Vries, M., 1975. *A morphological time scale for rivers.* Paper presented at the Proc. 16th Congr. IAHR, São Paulo, Brazil.

Dersseh, M. G., Kibret, A. A., Tilahun, S. A., Worqlul, A. W., Moges, M. A., Dagnew, D. C., Abebe, W. B., & Melesse, A. M., 2019. Potential of Water Hyacinth

Infestation on Lake Tana, Ethiopia: A Prediction Using a GIS-Based Multi-Criteria Technique. *Water, 11*(9), 1921. doi:10.3390/w11091921.

Dessie, M., Verhoest, N. E., Admasu, T., Pauwels, V. R., Poesen, J., Adgo, E., Deckers, J., & Nyssen, J., 2014. Effects of the floodplain on river discharge into Lake Tana (Ethiopia). *Journal of Hydrology, 519,* 699-710. doi:10.1016/j.jhydrol.2014.08.007.

Di Silvio, G., & Nones, M., 2014. Morphodynamic reaction of a schematic river to sediment input changes: Analytical approaches. *Geomorphology, 215,* 74-82.

Donchyts, G., Schellekens, J., Winsemius, H., Eisemann, E., & van de Giesen, N., 2016. A 30 m resolution surface water mask including estimation of positional and thematic differences using landsat 8, srtm and openstreetmap: a case study in the Murray-Darling Basin, Australia. *Remote Sensing, 8*(5), 386. doi:10.3390/rs8050386.

Duan, Z., Gao, H., & Ke, C., 2018. Estimation of Lake Outflow from the Poorly Gauged Lake Tana (Ethiopia) Using Satellite Remote Sensing Data. *Remote Sensing, 10*(7), 1060.

Duró, G., Crosato, A., & Tassi, P., 2016. Numerical study on river bar response to spatial variations of channel width. *Advances in water resources, 93(A),* 21-38. doi:10.1016/j.advwatres.2015.10.003.

Easton, Z., Fuka, D., White, E., Collick, A., Ashagre, B., McCartney, M., Awulachew, S., Ahmed, A., & Steenhuis, T., 2010. A multi basin SWAT model analysis of runoff and sedimentation in the Blue Nile, Ethiopia. *Hydrology and Earth System Sciences, 14*(10), 1827-1841.

Engelund, F., & Hansen, E., 1967. A monograph on sediment transport in alluvial streams. *Technical University of Denmark 0stervoldgade 10, Copenhagen K.*

ENTRO. 2010. *Flood risk mapping consultancy for pilot areas in Ethiopia. Final report to the Eastern Nile Technical Regional Office (ENTRO).* Addis Ababa, Ethiopia: ENTRO.

Exner, F. M., 1925. Uber die wechselwirkung zwischen wasser und geschiebe in flussen. *134*(2a), 165-204.

Feldman, A. D., 2000. *Hydrologic modeling system HEC-HMS: technical reference manual*: US Army Corps of Engineers, Hydrologic Engineering Center.

Fenton, J. D., 1992. Reservoir routing. *Hydrological Sciences Journal, 37*(3), 233-246.

Feyisa, G. L., Meilby, H., Fensholt, R., & Proud, S. R., 2014. Automated Water Extraction Index: A new technique for surface water mapping using Landsat imagery. *Remote sensing of environment, 140,* 23-35. doi:org/10.1016/j.rse.2013.08.029.

FitzHugh, T. W., & Vogel, R. M., 2011. The impact of dams on flood flows in the United States. *River Research Applications, 27*(10), 1192-1215.

Francis, I., & Aynalem, S., 2007. Bird surveys around Bahir Dar-Lake Tana IBA, Ethiopia. *Report of RSPB Scotland, Aberdeen, UK, Addis Ababa University, Ethiopia.*

Frings, R. M., 2015. Sand and gravel on the move: Human impacts on bed-material load along the Lower Rhine River. In *Geomorphic Approaches to Integrated Floodplain Management of Lowland Fluvial Systems in North America and Europe* (pp. 9-26): Springer.

Frings, R. M., Schüttrumpf, H., & Vollmer, S., 2011. Verification of porosity predictors for fluvial sand-gravel deposits. *Water Resources Research, 47*(7). doi:10.1029/2010WR009690 (W07525).

Gaeuman, D., Schmidt, J. C., & Wilcock, P. R., 2005. Complex channel responses to changes in stream flow and sediment supply on the lower Duchesne River, Utah. *Geomorphology, 64*(3), 185-206. doi:10.1016/j.geomorph.2004.06.007.

Garede, N. M., & Minale, A. S., 2014. Land use/cover dynamics in Ribb watershed, North Western Ethiopia. *Journal of Natural Sciences Research, 4*(16), 9-16.

Getahun, M., & G. Selassie, Y., 2017. Characterization, Classification and Mapping of Soils of Agricultural Landscape in Tana Basin, Amhara National Regional State, Ethiopia In K. Stave, G. Goshu, & S. Aynalem (Eds.), *Social and Ecological System Dynamics: Characteristics, Trends, and Integration in the Lake Tana Basin, Ethiopia* (pp. 93-115). Cham: Springer International Publishing.

Gilvear, D., & Winterbottom, S., 1992. Channel change and flood events since 1783 on the regulated River Tay, Scotland: Implications for flood hazard management. *Regulated Rivers: Research & Management, 7*(3), 247-260. doi:10.1002/rrr.3450070304.

Gilvear, D. J., 1993. River management and conservation issues on formerly braided river systems; the case of the River Tay, Scotland. *Geological Society, London, Special Publications, 75*(1), 231-240. doi:10.1144/GSL.SP.1993.075.01.14

Gilvear, D. J., 1999. Fluvial geomorphology and river engineering: Future roles utilizing a fluvial hydrosystems framework. *Geomorphology, 31*(1), 229-245. doi:10.1016/S0169-555X(99)00086-0.

Goshu, G., & Aynalem, S., 2017. Problem overview of the lake Tana basin. In K. Stave, G. Goshu, & S. Aynalem (Eds.), *Social and Ecological System Dynamics: Characteristics, Trends, and Integration in the Lake Tana Basin, Ethiopia* (pp. 9-23): Springer.

Graf, W. L., 2006. Downstream hydrologic and geomorphic effects of large dams on American Rivers. *Geomorphology, 79*(3), 336-360. doi:10.1016/j.geomorph.2006.06.022.

Grant, G. E., 2012. The Geomorphic Response of Gravel Bed Rivers to Dams: Perspectives and Prospects. *Gravel-Bed Rivers: Processes, Tools, Environments* (In: Church, M.; Biron, PM; Roy, AG, eds. ), 165-181. doi:10.1002/9781119952497.ch15.

Grant, G. E., Schmidt, J. C., & Lewis, S. L., 2003. A geological framework for interpreting downstream effects of dams on rivers. *Water Science and Application, 7*, 209-225. doi:10.1029/007WS13.

Gurnell, A. M., Downward, S. R., & Jones, R., 1994. Channel Planform Change on the River Dee Meanders, 1876-1992. *Regulated Rivers, 9*(4), 187.

Halwatura, D., & Najim, M., 2013. Application of the HEC-HMS model for runoff simulation in a tropical catchment. *Environmental modelling & software, 46*, 155-162.

Hamby, D., 1994. A review of techniques for parameter sensitivity analysis of environmental models. *Environmental Monitoring Assessment, 32*(2), 135-154. doi:10.1007/BF00547132.

Holly Jr, F. M., & Karim, M. F., 1986. Simulation of Missouri River bed degradation. *Hydraulic Engineering,* *112*(6), 497-516. doi:10.1061/(ASCE)0733-9429(1986)112:6(497).

Horritt, M., & Bates, P., 2002. Evaluation of 1D and 2D numerical models for predicting river flood inundation. *Journal of Hydrology, 268*(1-4), 87-99. doi:S0 02 2-1 69 4(0 2) 00 1 21-X.

Hu, X., Pant, R., Hall, J. W., Surminski, S., & Huang, J., 2019. Multi-Scale Assessment of the Economic Impacts of Flooding: Evidence from Firm to Macro-Level Analysis in the Chinese Manufacturing Sector. *Sustainability, 11*(7), 1933.

Huang, H. Q., & Nanson, G. C., 2000. Hydraulic geometry and maximum flow efficiency as products of the principle of least action. *Earth Surface Processes and Landforms: The Journal of the British Geomorphological Research Group, 25*(1), 1-16.

Hurni, H., Tato, K., & Zeleke, G., 2005. The implications of changes in population, land use, and land management for surface runoff in the Upper Nile Basin area of Ethiopia. *Mountain research and development, 25*(2), 147-154. doi:10.1659/0276-4741(2005)025[0147: TIOCIP] 2.0.CO; 2.

ICOLD. 1998. *World register of dams.* Paris: International Commission on Large Dams.

Jansen, P. P., Van Bendegom, L., De Vries, M., & Zenen, A., 1979. *Principles of River Engineering. The non-tidal alluvial river.* Delft, The Netherlands: Delftse Uitgevers Maatschappij B.V.

Ji, L., Zhang, L., & Wylie, B., 2009. Analysis of dynamic thresholds for the normalized difference water index. *Photogrammetric Engineering & Remote Sensing, 75*(11), 1307-1317.

Jiang, S. W., Haibier, A., & Wu, Y. X., 2013. Combined impacts of sand mining activities: The Nandu River downstream segment. *Advanced Materials Research, 671-674,* 3134-3137. doi:10.4028/www.scientific.net/AMR.671-674.3134.

Jin, H., Liang, R., Wang, Y., & Tumula, P., 2015. Flood-runoff in semi-arid and sub-humid regions, a case study: A simulation of Jianghe watershed in northern China. *Water, 7*(9), 5155-5172. doi:10.3390/w7095155.

Julian, D. W., Hickey, J. T., Fields, W. L., Ostadrahimi, L., Maher, K. M., Barker, T. G., Hatfield, C. L., Lutz, K., Marks, C. O., & Sandoval-Solis, S., 2016. Decision support system for water and environmental resources in the Connecticut River Basin. *Journal of Water Resources Planning and Management, 142*(1), 04015038.

Kadam, P., & Sen, D., 2012. Flood inundation simulation in Ajoy River using MIKE-FLOOD. *ISH Journal of Hydraulic Engineering, 18*(2), 129-141.

Kebede, S., 2012. *Groundwater in Ethiopia: features, numbers and opportunities*: Springer Science & Business Media.

Kebede, S., Admasu, G., & Travi, Y., 2011. Estimating ungauged catchment flows from Lake Tana floodplains, Ethiopia: an isotope hydrological approach. *Isotopes in Environmental and Health studies, 47*(1), 71-86.

Kebede, S., Travi, Y., Alemayehu, T., & Marc, V., 2006. Water balance of Lake Tana and its sensitivity to fluctuations in rainfall, Blue Nile basin, Ethiopia. *Journal of Hydrology, 316*(1), 233-247.

Khan, O., Mwelwa-Mutekenya, E., Crosato, A., & Zhou, Y., 2014. Effects of dam operation on downstream river morphology: The case of the Middle Zambezi

River. *Proceedings of the Institution of Civil Engineers - Water Management, 167*(10), 585-600. doi:0.1680/wama.13.00122.

Kinzel, P., & Runge, J., 2010. Summary of bed-sediment measurements along the Platte River, Nebraska, 1931–2009. *US Geol. Surv. Fact Sheet, 3087*(4).

Kirpich, Z., 1940. Time of concentration of small agricultural watersheds. *Civil engineering, 10*(6), 362.

Knebl, M., Yang, Z.-L., Hutchison, K., & Maidment, D. R., 2005. Regional scale flood modeling using NEXRAD rainfall, GIS, and HEC-HMS/RAS: a case study for the San Antonio River Basin Summer 2002 storm event. *Journal of Environmental Management, 75*(4), 325-336. doi:org/10.1016/j.jenvman.2004.11.024.

Knighton, D., 1998. *Fluvial Forms and Proceses*. New York: John Wiley and Sons, Inc.

Komi, K., Neal, J., Trigg, M. A., & Diekkrüger, B., 2017. Modelling of flood hazard extent in data sparse areas: a case study of the Oti River basin, West Africa. *Journal of Hydrology: Regional Studies, 10*, 122-132. doi:org/10.1016/j.ejrh.2017.03.001.

Kondolf, G. M., 1997. PROFILE: Hungry water: Effects of dams and gravel mining on river channels. *Environmental management, 7*(4), 303-325.

Kondolf, G. M., Gao, Y., Annandale, G. W., Morris, G. L., Jiang, E., Zhang, J., Cao, Y., Carling, P., Fu, K., Guo, Q., Hotchkiss, R., Peteuil, C., Sumi, T., Wang, H.-W., Wang, Z., Wei, Z., Wu, B., Wu, C., & Yang, C. T., 2014a. Sustainable sediment management in reservoirs and regulated rivers: Experiences from five continents. *Earth's Future, 2*(5), 256-280. doi:10.1002/2013EF000184.

Kondolf, G. M., Rubin, Z. K., & Minear, J. T., 2014b. Dams on the Mekong: Cumulative sediment starvation. *Water Resources Research, 50*(6), 5158-5169. doi:10.1002/2013WR014651.

Lagasse, P. F., Zevenbergen, L., Spitz, W., & Thorne, C., 2004. *Methodology for predicting channel migration*. Transportation Research Board, National Research Council.

Lanzoni, S., Luchi, R., & Pittaluga, M. B., 2015. Modeling the morphodynamic equilibrium of an intermediate reach of the Po River (Italy). *Advances in water resources, 81*, 95-102.

Latapie, A., Camenen, B., Rodrigues, S., Paquier, A., Bouchard, J., & Moatar, F., 2014. Assessing channel response of a long river influenced by human disturbance. *Catena, 121*, 1-12. doi:10.1016/J.catena.2014.04.017.

Leggesse, E. S., & Beyene, B. S., 2017. Hydrology of Lake Tana Basin. In K. Stave, G. Goshu, & S. Aynalem (Eds.), *Social and Ecological System Dynamics: Characteristics, Trends, and Integration in the Lake Tana Basin, Ethiopia* (pp. 117-126): Springer.

Legleiter, C. J., 2014. Downstream Effects of Recent Reservoir Development on the Morphodynamics of a Meandering Channel: Savery Creek, Wyoming, USA. *River Research and Applications*. doi:10.1002/rra.2824.

Leon, A. S., Kanashiro, E. A., Valverde, R., & Sridhar, V., 2014. Dynamic framework for intelligent control of river flooding: Case study. *Journal of Water Resources Planning and Management, 140*(2), 258-268.

Leopold, L. B., & Wolman, M. G., 1957. *River channel patterns: braided, meandering, and straight*: US Government Printing Office.

Leopold, L. B., Wolman, M. G., & Miller, J. P., 1964. *Fluvial processes in Geomorphology.* San Francisco, USA: W.H. Freeman and Company.

Li, F.-F., Liu, C.-M., Wu, Z.-G., & Qiu, J., 2020. Balancing Ecological Requirements and Power Generation in Reservoir Operation in Fish Spawning Seasons. *Journal of Water Resources Planning and Management, 146*(9), 04020074.

Li, L., Lu, X., & Chen, Z., 2007. River channel change during the last 50 years in the Middle Yangtze River, the Jianli Reach. *Geomorphology, 85*(3), 185-196. doi:10.1016/j.geomorph.2006.03.035.

Li, S., Li, Y., Yuan, J., Zhang, W., Chai, Y., & Ren, J., 2018. The impacts of the Three Gorges Dam upon dynamic adjustment mode alterations in the Jingjiang reach of the Yangtze River, China. *Geomorphology, 318*, 230-239.

Li, W., Du, Z., Ling, F., Zhou, D., Wang, H., Gui, Y., Sun, B., & Zhang, X., 2013. A comparison of land surface water mapping using the normalized difference water index from TM, ETM+ and ALI. *Remote Sensing, 5*(11), 5530-5549. doi:10.3390/rs5115530.

Li, W., Wang, Z., de Vriend, H. J., & van Maren, D. S., 2014. Long-Term Effects of Water Diversions on the Longitudinal Flow and Bed Profiles. *Hydraulic Engineering, 140*(6), 04014021. doi:10.1061/(ASCE)HY.1943-7900.0000856.

Limbu, S. M., 2020. The effects of on-farm produced feeds on growth, survival, yield and feed cost of juvenile African sharptooth catfish (Clarias gariepinus). *Aquaculture and Fisheries, 5*(1), 58-64.

Liu, B. M., Collick, A. S., Zeleke, G., Adgo, E., Easton, Z. M., & Steenhuis, T. S., 2008. Rainfall-discharge relationships for a monsoonal climate in the Ethiopian highlands. *Hydrological Processes: An International Journal, 22*(7), 1059-1067. doi:org/10.1002/hyp.7022.

Lobera, G., Besné, P., Vericat, D., López-Tarazón, J. A., Tena, A., Aristi, I., Díez, J. R., Ibisate, A., Larrañaga, A., & Elosegi, A., 2015. Geomorphic status of regulated rivers in the Iberian Peninsula. *Science of the Total Environment, 508*, 101-114.

Lytle, D. A., & Poff, N. L., 2004. Adaptation to natural flow regimes. *Trends in ecology & evolution, 19*(2), 94-100.

Magilligan, F. J., & Nislow, K. H., 2005. Changes in hydrologic regime by dams. *Geomorphology, 71*(1), 61-78. doi:10.1016/j.geomorph.2004.08.017.

Marcinkowski, P., & Grygoruk, M., 2017. Long-term downstream effects of a dam on a lowland river flow regime: Case study of the Upper Narew. *Water, 9*(10), 783.

Marston, R. A., Girel, J., Pautou, G., Piegay, H., Bravard, J.-P., & Arneson, C., 1995. Channel metamorphosis, floodplain disturbance, and vegetation development: Ain River, France. *Geomorphology, 13*(1), 121-131. doi:10.1016/0169-555X (95)00066-E.

McCartney, M., Alemayehu, T., Shiferaw, A., & Awulachew, S., 2010. *Evaluation of current and future water resources development in the Lake Tana Basin, Ethiopia* (Vol. 134): IWMI.

Mei, X., Van Gelder, P., Dai, Z., & Tang, Z., 2017. Impact of dams on flood occurrence of selected rivers in the United States. *Frontiers of Earth Science, 11*(2), 268-282. doi:10.1007/s11707-016-0592-1.

Mengistu, A. A., Aragaw, C., Mengist, M., & Goshu, G., 2017. The fish and the fisheries of Lake Tana. In K. Stave, G. Goshu, & S. Aynalem (Eds.), *Social and Ecological*

*System Dynamics: Characteristics, Trends, and Integration in the Lake Tana Basin, Ethiopia* (pp. 157-177): Springer.

Merwade, V., Olivera, F., Arabi, M., & Edleman, S., 2008. Uncertainty in flood inundation mapping: current issues and future directions. *Journal of Hydrologic Engineering, 13*(7), 608-620.

Meyer-Peter, E., & Müller, R., 1948. *Formulas for bed-load transport.* Paper presented at the IAHR 2nd meeting, Stockholm, Appendix 2.

Minale, A. S., & Belete, W., 2017. Land use distribution and change in lake Tana sub basin. In *Social and Ecological System Dynamics* (pp. 357-373): Springer.

Mitsch, W. J., 2005. Invitational at the Olentangy River Wetland Research Park. *Wetl Creation Restor Conserv State Sci, 24*, 243-251.

Mohammadi, S., Nazariha, M., & Mehrdadi, N., 2014. Flood damage estimate (quantity), using HEC-FDA model. Case study: the Neka river. *Procedia Engineering, 70*, 1173-1182. doi:org/10.1016/j.proeng.2014.02.130.

Mohammed, I., & Mengist, M., 2019. Status, threats and management of wetlands in the Lake Tana sub-basin: a review. *Journal of Agriculture Environmental Sciences, 3*(2), 23-45.

Moriasi, D. N., Arnold, J. G., Van Liew, M. W., Bingner, R. L., Harmel, R. D., & Veith, T. L., 2007. Model evaluation guidelines for systematic quantification of accuracy in watershed simulations. *Transactions of the ASABE, 50*(3), 885-900.

MoWR. 2002. *Main Report: Water Sector Development Program.* Ethiopia: MoWR.

Mulatu, C. A., Boeriu, P., & Bekele, S., 2012. Analysis of reservoir sedimentation process using empirical and mathematical method: case study—Koga Irrigation and Watershed Management Project; Ethiopia. *Nile Basin Water Science Engineering Journal, 5*(1), 1-13.

Mulatu, C. A., Crosato, A., Moges, M. M., Langendoen, E. J., & McClain, M., 2018. Morphodynamic Trends of the Ribb River, Ethiopia, Prior to Dam Construction. *Geosciences, 8*, 255. doi:10.3390/geosciences8070255.

Mulatu, C. A., Crosato, A., Moges, M. M., Langendoen, E. J., & McClain, M., 2020. Long-term effects of dam operations for water supply to irrigation on downstream river reaches. The case of the Ribb River, Ethiopia. *The International Journal of River Basin Management.* doi:10.1080/15715124.2020.1750421.

Mulatu, C. A., Crosato, A., & Mynett, A., 2017. *Analysis of Ribb River channel migration: Upper Blue Nile, Ethiopia.* Paper presented at the Netherlands Centre for River Studies., Wageningen, Netherlands.

Mundt, F., 2011. *Wetlands around Lake Tana: a landscape and avifaunistic study.* (M.Sc.), Universitat Greifswald, Germany.

Negash, A., Eshete, D., & Jacobus, V., 2011. Assessment of the ecological status and threats of Welala and Shesher Wetlands, Lake Tana sub-basin (Ethiopia). *Journal of Water Resource and Protection, 2011.* doi:10.4236/jwarp.2011.37064.

Nelson, J. M., Shimizu, Y., Abe, T., Asahi, K., Gamou, M., Inoue, T., Iwasaki, T., Kakinuma, T., Kawamura, S., & Kimura, I., 2016. The international river interface cooperative: Public domain flow and morphodynamics software for education and applications. *Advances in water resources, 93*, 62-74. doi:10.1016/j.advwatres.2015.09.017.

121

Nigate, F., 2019. *Investigating the hydrogeological system of the Lake Tana basin, in the northwestern highlands of Ethiopia (the Upper Blue Nile).* (PhD), Ghent University, Belgium.

Nigate, F., Ayenew, T., Belete, W., & Walraevens, K., 2017. Overview of the Hydrogeology and Groundwater Occurrence in the Lake Tana Basin, Upper Blue Nile River Basin. In K. Stave, G. Goshu, & S. Aynalem (Eds.), *Social and Ecological System Dynamics: Characteristics, Trends, and Integration in the Lake Tana Basin, Ethiopia* (pp. 77-91): Springer.

Nigate, F., Camp, M. V., Yenehun, A., Belay, A. S., & Walraevens, K., 2020. Recharge–Discharge Relations of Groundwater in Volcanic Terrain of Semi-Humid Tropical Highlands of Ethiopia: The Case of Infranz Springs, in the Upper Blue Nile. *Water, 12*(3), 853.

Nigate, F., Van Camp, M., Kebede, S., & Walraevens, K., 2016. Hydrologic interconnection between the volcanic aquifer and springs, Lake Tana basin on the Upper Blue Nile. *Journal of African Earth Sciences, 121*, 154-167. doi:10.1016/j.jafrearsci.2016.05.015.

Nones, M., Guerrero, M., & Ronco, P., 2014. Opportunities from low-resolution modelling of river morphology in remote parts of the world. *Earth Surface Dynamics, 2*(1), 9-19. doi:10.5194/esurf-2-9-2014.

Nones, M., Varrani, A., Franzoia, M., & Di Silvio, G., 2019. Assessing quasi-equilibrium fining and concavity of present rivers: A modelling approach. *Catena, 181*, 104073.

Ohl, C. A., & Tapsell, S., 2000. Flooding and human health: the dangers posed are not always obvious. *Brit. Med. J., 321*, 1167–1168. doi:10.1136/bmj.321.7270.1167.

Omer, A., Ali, Y., Roelvink, J., Dastgheib, A., Paron, P., & Crosato, A., 2015. Modelling of sedimentation processes inside Roseires Reservoir (Sudan). *Earth Surface Dynamics, 3*(2), 223-238.

Opperman, J. J., Galloway, G. E., Fargione, J., Mount, J. F., Richter, B. D., & Secchi, S., 2009. Sustainable floodplains through large-scale reconnection to rivers. *Science, 326*(5959), 1487-1488. doi:10.1126/science.1178256.

Osterkamp, W., Scott, M. L., & Auble, G. T., 1998. Downstream effects of dams on channel geometry and bottomland vegetation: regional patterns in the Great Plains. *Wetlands, 18*(4), 619-633.

Otsu, N., 1979. A threshold selection method from gray-level histograms. *IEEE transactions on systems, man, cybernetics, 9*(1), 62-66. doi:10.1109/TSMC.1979.4310076.

Oubennaceur, K., Chokmani, K., Nastev, M., Tanguy, M., & Raymond, S., 2018. Uncertainty analysis of a two-dimensional hydraulic model. *Water, 10*(3), 272.

Pappenberger, F., Beven, K., Horritt, M., & Blazkova, S., 2005. Uncertainty in the calibration of effective roughness parameters in HEC-RAS using inundation and downstream level observations. *Journal of Hydrology, 302*(1-4), 46-69.

Parker, G., Wilcock, P. R., Paola, C., Dietrich, W. E., & Pitlick, J., 2007. Physical basis for quasi-universal relations describing bankfull hydraulic geometry of single-thread gravel bed rivers. *Journal of Geophysical Research: Earth Surface, 112*(F4).

Patro, S., Chatterjee, C., Mohanty, S., Singh, R., & Raghuwanshi, N., 2009. Flood inundation modeling using MIKE FLOOD and remote sensing data. *Journal of the Indian Society of Remote Sensing, 37*(1), 107-118.

Petts, G. E., 1979. Complex response of river channel morphology subsequent to reservoir construction. *Progress in Physical Geography, 3*(3), 329-362. doi:10.1177/030913337900300302.

Petts, G. E., 1980. Long-term consequences of upstream impoundment. *Environmental Conservation, 7*(04), 325-332.

Petts, G. E., & Gurnell, A. M., 2005. Dams and geomorphology: research progress and future directions. *Geomorphology, 71*(1), 27-47. doi:10.1016/j.geomorph.2004.02.015.

Phillips, J. D., Slattery, M. C., & Musselman, Z. A., 2005. Channel adjustments of the lower Trinity River, Texas, downstream of Livingston Dam. *Earth Surface Processes and Landforms: The Journal of the British Geomorphological Research Group, 30*(11), 1419-1439.

Pilon, P. J., 2002. *Guidelines for reducing flood losses*. Retrieved from United Nations International Strategy for Disaster Reduction (UNISDR).

Pinos, J., & Timbe, L., 2019. Performance assessment of two-dimensional hydraulic models for generation of flood inundation maps in mountain river basins. *Water Science Engineering, 12*(1), 11-18. doi:doi.org/10.1016/j.wse.2019.03.001.

Poff, N., Allan, J., Bain, M., Karr, J., Prestegaard, K., Richter, B., Sparks, R., & Stromberg, J., 1997. The natural flow regime, a paradigm for river conservation and restoration. *Bioscience, 47*, 769–784.

Popescu, I., Jonoski, A., Van Andel, S., Onyari, E., & Moya Quiroga, V., 2010. Integrated modelling for flood risk mitigation in Romania: case study of the Timis–Bega river basin. *International Journal of River Basin Management, 8*(3-4), 269-280. doi:10.1080/15715124.2010.512550.

Poppe, L., Frankl, A., Poesen, J., Admasu, T., Dessie, M., Adgo, E., Deckers, J., & Nyssen, J., 2013. Geomorphology of the Lake Tana basin, Ethiopia. *Journal of Maps, 9*(3), 431-437.

Prave, A., Bates, C. R., Donaldson, C. H., Toland, H., Condon, D., Mark, D., & Raub, T. D., 2016. Geology and geochronology of the Tana Basin, Ethiopia: LIP volcanism, super eruptions and Eocene–Oligocene environmental change. *Earth and Planetary Science Letters, 443*, 1-8.

Qicai, L., 2011. Influence of dams on river ecosystem and its countermeasures. *Journal of Water Resource and Protection*. doi:10.4236/jwarp.2011.31007.

Quirogaa, V. M., Kurea, S., Udoa, K., & Manoa, A., 2016. Application of 2D numerical simulation for the analysis of the February 2014 Bolivian Amazonia flood: Application of the new HEC-RAS version 5. *Ribagua, 3*(1), 25-33.

Rauf, A.-u., & Ghumman, A. R., 2018. Impact assessment of rainfall-runoff simulations on the flow duration curve of the Upper Indus River—a comparison of data-driven and hydrologic models. *Water, 10*(7), 876.

Rendon, S. H., Ashworth, C. E., & Smith, S. J., 2012. *Dam-breach Analysis and Flood-inundation Mapping for Lakes Ellsworth and Lawtonka Near Lawton, Oklahoma*: US Department of Interior, US Geological Survey.

Richter, B. D., Mathews, R., Harrison, D. L., & Wigington, R., 2003. Ecologically sustainable water management: managing river flows for ecological integrity. *Ecological applications, 13*(1), 206-224.

Ronco, P., Fasolato, G., Nones, M., & Di Silvio, G., 2010. Morphological effects of damming on lower Zambezi River. *Geomorphology, 115*(1), 43-55.

Rubin, Z. K., Kondolf, G. M., & Carling, P. A., 2015. Anticipated geomorphic impacts from Mekong basin dam construction. *International Journal of River Basin Management, 13*(1), 105-121.

Sanyal, J., 2017. Predicting possible effects of dams on downstream river bed changes of a Himalayan river with morphodynamic modelling. *Quaternary International, 453*, 48-62.

Sayama, T., Ozawa, G., Kawakami, T., Nabesaka, S., & Fukami, K., 2012. Rainfall–runoff–inundation analysis of the 2010 Pakistan flood in the Kabul River basin. *Hydrological Sciences Journal, 57*(2), 298-312.

Scharffenberg, W. A., & Fleming, M. J., 2016. *Hydrologic modeling system HEC-HMS: user's manual*: US Army Corps of Engineers, Hydrologic Engineering Center.

Schmidt, J. C., & Wilcock, P. R., 2008. Metrics for assessing the downstream effects of dams. *Water Resources Research, 44*(4). doi:10.1029/2006WR005092.

Setegn, S. G., Rayner, D., Melesse, A. M., Dargahi, B., & Srinivasan, R., 2011. Impact of climate change on the hydroclimatology of Lake Tana Basin, Ethiopia. *Water Resources Research, 47*(4).

Setegn, S. G., Srinivasan, R., & Dargahi, B., 2008. Hydrological modelling in the Lake Tana Basin, Ethiopia using SWAT model. *The Open Hydrology Journal, 2*(2008), 49-62.

Shamsudin, S., Dan'azumi, S., & Ab Rahman, A., 2011. Uncertainty analysis of HEC-HMS model parameters using Monte Carlo simulation. *International Journal of Modelling and Simulation, 31*(4), 279-286.

Sharafati, A., Khazaei, M. R., Nashwan, M. S., Al-Ansari, N., Yaseen, Z. M., & Shahid, S., 2020. Assessing the Uncertainty Associated with Flood Features due to Variability of Rainfall and Hydrological Parameters. *Advances in Civil Engineering, 2020*.

Shaw, E. M., Beven, K. J., Chappell, N. A., & Lamb, R., 2010. *Hydrology in practice.*: CRC Press.

Shields Jr, F. D., Simon, A., & Steffen, L., 2000. Reservoir effects on downstream river channel migration. *Environmental Conservation, 27*(1), 54-66.

SMEC. 2008a. *Hydrological study of the Tana-Beles Sub-basin: Main Report.* (5089018). Australia: Snowy Mountains Engineering Corporation (SMEC) International Pty Ltd.

SMEC. 2008b. *Hydrological study of the Tana-Beles Sub-basin: Surface water investigation.* Australia: Snowy Mountains Engineering Corporation (SMEC) International Pty Ltd.

Surian, N., 1999. Channel changes due to river regulation: the case of the Piave River, Italy. *Earth Surface Processes and Landforms, 24*(12), 1135-1151. doi:10.1002/(SICI)1096-9837(199911)24:123.3.CO;2-6.

Surian, N., Rinaldi, M., Pellegrini, L., Audisio, C., Maraga, F., Teruggi, L., Turitto, O., & Ziliani, L., 2009. Channel adjustments in northern and central Italy over the last 200 years. *Management and restoration of fluvial systems with broad historical* changes and human impacts: geological society of America Special Paper, 451, 83-95.

Svetlana, D., Radovan, D., & Ján, D., 2015. The economic impact of floods and their importance in different regions of the world with emphasis on Europe. *Procedia Economics and Finance, 34*, 649-655. doi:10.1016/S2212-5671(15)01681-0.

Syvitski, J. P., Vörösmarty, C. J., Kettner, A. J., & Green, P., 2005. Impact of humans on the flux of terrestrial sediment to the global coastal ocean. *Science, 308*(5720), 376-380.

Talukdar, S., & Pal, S., 2019. Effects of damming on the hydrological regime of Punarbhaba river basin wetlands. *Ecological Engineering, 135*, 61-74.

Tang, Z., Li, Y., Gu, Y., Jiang, W., Xue, Y., Hu, Q., LaGrange, T., Bishop, A., Drahota, J., & Li, R., 2016. Assessing Nebraska playa wetland inundation status during 1985–2015 using Landsat data and Google Earth Engine. *Environmental Monitoring Assessment, 188*(12), 654. doi:10.1007/s10661-016-5664-x.

Tassew, B. G., Belete, M. A., & Miegel, K., 2019. Application of HEC-HMS model for flow simulation in the Lake Tana basin: The case of Gilgel Abay catchment, upper Blue Nile basin, Ethiopia. *Hydrology, 6*(1), 21.

Tefera, B., 2017. Water-Induced Shift of Farming Systems and Value Addition in Lake Tana Sub-basin: The Case of Rice Production and Marketing in Fogera District, Northwestern Ethiopia. In K. Stave, G. Goshu, & S. Aynalem (Eds.), *Social and Ecological System Dynamics: Characteristics, Trends, and Integration in the Lake Tana Basin, Ethiopia* (pp. 545-562): Springer.

Tekleab, S., Mohamed, Y., & Uhlenbrook, S., 2013. Hydro-climatic trends in the Abay/Upper Blue Nile basin, Ethiopia. *Physics and Chemistry of the Earth, Parts A/B/C, 61*, 32-42. doi:10.1016/j.pce.2013.04.017.

Teng, J., Jakeman, A. J., Vaze, J., Croke, B. F., Dutta, D., & Kim, S., 2017. Flood inundation modelling: A review of methods, recent advances and uncertainty analysis. *Environmental modelling software, 90*, 201-216. doi:org/10.1016/j.envsoft.2017.01.006.

Teruggi, L., & Rinaldi, M., 2009. *Analysis of planimetric channel changes along the Cecina River (central Italy)*. Paper presented at the International proceedings of 27th IAS meeting of sedimentology. Medimond, Bologna, Italy.

Tesemma, Z. K., Mohamed, Y. A., & Steenhuis, T. S., 2010. Trends in rainfall and runoff in the Blue Nile Basin: 1964–2003. *Hydrological Processes, 24*(25), 3747-3758. doi:10.1002/hyp.7893.

Thoms, M., & Walker, K., 1993. Channel changes associated with two adjacent weirs on a regulated lowland alluvial river. *River Research and Applications, 8*(3), 271-284. doi:10.1002/rrr.3450080306.

Thornton, E. B., Sallenger, A., Sesto, J. C., Egley, L., McGee, T., & Parsons, R., 2006. Sand mining impacts on long-term dune erosion in Southern Monterey Bay. *Marine Geology, 229*(1), 45-58. doi:10.1016/j.margeo.2006.02.005.

Uhlenbrook, S., Mohamed, Y., & Gragne, A., 2010. Analyzing catchment behavior through catchment modeling in the Gilgel Abay, upper Blue Nile River basin, Ethiopia. *Hydrology and Earth System Sciences, 14*(10), 2153.

Van den Berg, J. H., 1995. Prediction of alluvial channel pattern of perennial rivers. *Geomorphology, 12*(4), 259-279. doi:10.1016/0169-555X (95)00014-V.

Van der Zwet, J., 2012. *The creation of a reservoir in the White Volta River, Ghana: An analysis of the impact on river morphology.* (M. Sc.), TU Delft, Delft, The Netherlands.

Vargas-Luna, A., Crosato, A., Byishimo, P., & Uijttewaal, W. S., 2018. Impact of flow variability and sediment characteristics on channel width evolution in laboratory streams. *Journal of Hydraulic Research,* 1-11. doi:org/10.1080/00221686.2018.1434836.

Varrani, A., Nones, M., & Gupana, R., 2019. Long-term modelling of fluvial systems at the watershed scale: examples from three case studies. *Journal of Hydrology, 574,* 1042-1052.

Vijverberg, J., Sibbing, F. A., & Dejen, E., 2009. Lake Tana: Source of the Blue Nile. In *The Nile* (pp. 163-192): Springer.

Vojtek, M., Petroselli, A., Vojteková, J., & Asgharinia, S., 2019. Flood inundation mapping in small and ungauged basins: sensitivity analysis using the EBA4SUB and HEC-RAS modeling approach. *Hydrology Research.* doi:10.2166/nh.2019.163.

Ward, J., Tockner, K., Arscott, D. B., & Claret, C., 2002. Riverine landscape diversity. *Freshwater Biology, 47*(4), 517-539. doi:org/10.1046/j.1365-2427.2002.00893.x.

WBISPP. 2002. *Report on natural grazing lands and livestock feed resources, Amhara National Regional State, Woody Biomass Inventory and Strategic Planning Project (WBISPP).* Addis Ababa.

Weldegerima, T. M., Zeleke, T. T., Birhanu, B. S., Zaitchik, B. F., & Fetene, Z. A., 2018. Analysis of rainfall trends and its relationship with SST signals in the Lake Tana Basin, Ethiopia. *Advances in Meteorology, 2018.*

Wild, T. B., Reed, P. M., Loucks, D. P., Mallen-Cooper, M., & Jensen, E. D., 2019. Balancing hydropower development and ecological impacts in the Mekong: Tradeoffs for sambor mega dam. *Journal of Water Resources Planning and Management, 145*(2), 05018019.

Wilkerson, G. V., & Parker, G., 2010. Physical basis for quasi-universal relationships describing bankfull hydraulic geometry of sand-bed rivers. *Journal of Hydraulic Engineering, 137*(7), 739-753.

Williams, G. P., 1978. Bank-full discharge of rivers. *Water Resources Research, 14*(6), 1141-1154. doi:10.1029/WR014I006P01141.

Williams, G. P., & Wolman, M. G., 1984. *Downstream effects of dams on alluvial rivers.* Washington, D.C. 20402: U.S. Government Printing Office.

Winterbottom, S. J., 2000. Medium and short-term channel planform changes on the Rivers Tay and Tummel, Scotland. *Geomorphology, 34*(3), 195-208.

Wondie, A., 2018. Ecological conditions and ecosystem services of wetlands in the Lake Tana Area, Ethiopia. *Ecohydrology and Hydrobiology, 18*(2), 231-244. doi:org/10.1016/j.ecohyd.2018.02.002.

Wong, M., & Parker, G., 2006. Reanalysis and correction of bed-load relation of Meyer-Peter and Müller using their own database. *Hydraulic Engineering, 132*(11), 1159-1168. doi:10.1061/ (ASCE) 0733-9429(2006)132:11(1159).

Worqlul, A. W., Collick, A. S., Rossiter, D. G., Langan, S., & Steenhuis, T. S., 2015. Assessment of surface water irrigation potential in the Ethiopian highlands: The Lake Tana Basin. *Catena, 129,* 76-85.

Wu, B. S., Zheng, S., & Thorne, C. R., 2012. A general framework for using the rate law to simulate morphological response to disturbance in the fluvial system. *Prog. Phys. Geogr.* , *36*(5), 575–597.

WWDSE, & TAHAL. 2007. *Ribb Dam Hydrological Study (Final Report).* Addis Ababa, Ethiopia: Water Works Design and Supervision Enterprise (WWDSE) and TAHAL Consulting Engineers Ltd.

Xu, H., 2006. Modification of normalised difference water index (NDWI) to enhance open water features in remotely sensed imagery. *International journal of remote sensing, 27*(14), 3025-3033. doi:org/10.1080/01431160600589179.

Yang, S. L., Milliman, J. D., Li, P., & Xu, K., 2011. 50,000 dams later: Erosion of the Yangtze River and its delta. *Global and Planetary Change, 75*(1), 14-20. doi:10.1016/j.gloplacha.2010.09.006.

Yang, Y.-C. E., & Cai, X., 2011. Reservoir reoperation for fish ecosystem restoration using daily inflows—Case study of Lake Shelbyville. *Journal of Water Resources Planning and Management, 137*(6), 470-480.

Zeleke, G., & Hurni, H., 2001. Implications of land use and land cover dynamics for mountain resource degradation in the Northwestern Ethiopian highlands. *Mountain research and development, 21*(2), 184-191.

Zelelew, D., & Melesse, A., 2018. Applicability of a spatially semi-distributed hydrological model for watershed scale runoff estimation in Northwest Ethiopia. *Water, 10*(7), 923. doi:10.3390/w10070923.

Zelelew, D. G., & Melesse, A. M., 2018. Applicability of a spatially semi-distributed hydrological model for watershed scale runoff estimation in Northwest Ethiopia. *Water, 10*(7), 923.

Zhang, G., & Savenije, H., 2005. Rainfall-runoff modelling in a catchment with a complex groundwater flow system: application of the Representative Elementary Watershed (REW) approach. *Hydrology and Earth System Sciences*(9), 243–261. doi:.org/10.5194/hess-9-243-2005.

Zhou, M., Xia, J., Deng, S., Lu, J., & Lin, F., 2018. Channel adjustments in a gravel-sand bed reach owing to upstream damming. *Global planetary change, 170*, 213-220.

# LIST OF ACRONYMS

| | |
|---|---|
| 1D | One dimensional |
| 1D2D | One dimensional Two Dimensional |
| 2D | Two dimensional |
| 3D | Three dimensional |
| ASTER DEM | Advanced Spaceborne Thermal Emission and Reflection Radiometer Digital Elevation Model |
| DEM | Digital Elevation Model |
| EELPA | Ethiopian Electricity Light and Power Authority |
| EEPCo | Ethiopian Electricity Power Corporation |
| ENTRO | Eastern Nile Technical Regional Office |
| ENVI | Environment for Visualizing Images |
| ESC | Ethiopian Sugar Corporation |
| GCPs | Ground Control Points |
| GEE | Google Earth Engine |
| GERD | Grand Ethiopian Renaissance Dam |
| GIS | Geographical Information System |
| HEC-GeoHMS | Hydrologic Engineering Center-Geospatial Hydrologic Modeling |
| HEC-HMS | Hydrologic Engineering Center-Hydrologic Modelling System |
| HEC-RAS | Hydrologic Engineering Center-River Analysis System |
| ICOLD | International Commission on Large Dams |
| ITCZ | Inter-Tropical Convergence Zone |
| LULC | Land Use and Land Cover |
| m.a.s.l | Meters above sea level |
| MCE | Metaferia Consultancy Engineers |
| M | Mean of Modified Normalized Difference Water Index values |
| Mw | Mean pixel values for the water classes |
| Mnw | Mean pixel values for the non-water classes |
| MNDWI | Modified Normalized Difference Water Index |
| MoWIE | Ministry of Water, Irrigation and Energy |
| MoWR | Ministry of Water Resources |
| NMA | National Meteorological Agency |
| NPL | Normal Pool Level of the dam |
| NSE | Nash-Sutcliff Efficiency |
| $P_w$ | Probability of the pixel being in the water classes |
| $P_{nw}$ | Probability of the pixel being in the non-water classes |
| $R^2$ | coefficient of determination |
| RMSE | Root Mean Square Error |

| | |
|---|---|
| SMEC | Snowy Mountains Engineering Corporation |
| SPOT | Satellite Pour l'Observation de la Terre in French, meaning Satellite for observation of Earth |
| SRTM | Shuttle Radar Topographic Mission |
| TM | Thematic Mapper (Landsat satellite images) |
| TP | Thiessen Polygon |
| UBNB | Upper Blue Nile Basin |
| UH | Unit Hydrograph |
| UNESCO | United Nations Educational, Scientific and Cultural Organization |
| USACE | United States Army Corps of Engineers |
| USD | United States Dollar |
| WWDSE | Water Works Design and Supervision Enterprise |

# LIST OF TABLES

# LIST OF FIGURES

# ABOUT THE AUTHOR

Chalachew Abebe Mulatu was born and raised in Dangila, Ethiopia. He obtained his bachelor's degree in Hydraulic Engineering from Arba Minch Water Technology Institute (AWTI), currently, Arba Minch University in 2000. He has been working in the Amhara National Regional State Water Resources Development Bureau (ANRS-BoWRD), Commission for Sustainable Agriculture and Environmental Rehabilitation in Amhara Region, and Amhara Water Works Construction Enterprise under different positions since 2005. He participated mainly in the planning, identification, study, and design of small and medium scale dams, diversion weirs, intake structures, and farm structures; prepare the bill of quantities and estimate costs of the projects, construction plan, and participate in the construction supervision of irrigation projects. He received his M.Sc. degree in Water Science and Engineering, specialization Hydraulic Engineering and River Basin Development, from UNESCO-IHE, Delft, the Netherlands, in April 2007.

Back to his home country, he has been worked as a design irrigation engineer in ANRS-BoWRD and as an Irrigation and Drainage Expert in Sustainable Water Harvesting and Institutional Strengthening in Amhara (SWHISA) (a project funded by Canadian International Development Agency (CIDA)). In May 2009, he joined the Faculty of Civil and Water Resources Engineering, Bahir Dar Institute of Technology, Bahir Dar University in a position of lecturer. In 2015, he joined IHE-Delft receiving a scholarship from the Netherlands Fellowship Program (NFP) for his Ph.D. His research work focuses on the effects of dam construction on the downstream river system, focusing on the Ribb River, Ethiopia.

## Journal publications

**Mulatu, C. A.,** Crosato, A., M., Langendoen, E. J., Moges, M., & McClain, M. (2021). Flood attenuation and its ecological impact due to upstream dam construction, the case of Ribb Dam, Upper Blue Nile Basin, Ethiopia. Journal of Applied Water Engineering and Research. doi.org/10.1080/23249676.2021.1961618.

**Mulatu, C. A.,** Crosato, A., M., Langendoen, E. J., Moges, M., & McClain, M. (2020). Long-term effects of dam operations for water supply to irrigation on downstream river reaches. The case of the Ribb River, Ethiopia. The International Journal of River Basin Management. doi:10.1080/15715124.2020.1750421.

**Mulatu, C. A.,** Crosato, A., Moges, M. M., Langendoen, E. J., & McClain, M. (2018). Morphodynamic Trends of the Ribb River, Ethiopia, Prior to Dam Construction. Geosciences, 8, p 255. Doi: 10.3390/geosciences8070255.

**Mulatu, C. A.,** Boeriu, P., & Bekele, S. (2012). Analysis of reservoir sedimentation process using empirical and mathematical method: case study—Koga Irrigation and

Watershed Management Project; Ethiopia. Nile Basin Water Science Engineering Journal, 5(1), pp.1-13.

## Conference proceedings

**Mulatu**, C. A., Crosato, A., Langendoen, E. J., Moges, M. M., & McClain, M. (2020). Flood Attenuation and Ecological Impact of the Ribb Dam Construction on the Fogera Plain. PhD Symposium on Collaboration for sustainability, IHE-Delft, October 7-8, 2020, Delft, The Netherlands.

Abera, W., Haregeweyn, N., Dile, Y., Fenta, A. A., Berihun, M. L., Demissie, B., **Mulatu, C. A.**, Nigussie, T. A., Billi, P., & Meaza, H. (2020). Scientific Misconduct and Partisan Research on the Stability of the Grand Ethiopian Renaissance Dam: A Critical Review of a Contribution to Environmental Remote Sensing in Egypt (Springer, 2020). Paper presented at the 2020 International Conference on the Nile and Grand Ethiopian Renaissance Dam: Science, Conflict Resolution and Cooperation, August 20-21, 2020 Miami, FL, USA (Virtual).

**Mulatu**, C. A., Crosato, A., Langendoen, E. J., Moges, M. M., & McClain, M. E. (2019). Quick Assessment Method to Establish Long-Term Effects of Dam Operations on Downstream Rivers. AGUFM, 2019, San Francisco, USA, EP33D-2386.

**Mulatu**, C. A., Crosato, A., Langendoen, E. J., Moges, M. M., & McClain, M. (2018). Morphodynamic Trends of the Ribb River, Ethiopia, Prior to Dam Construction. PhD Symposium on Nature for Water: Overcoming Water Challenges with Sustainable Solutions, UNESCO-IHE, October 28-29, 2018, Delft, The Netherlands.

**Mulatu**, C. A., Crosato, A., & Mynett, A. (2017). Analysis of Ribb River channel migration: Upper Blue Nile, Ethiopia. Netherlands Centre for River Studies. Wageningen University, Netherlands. In the NCR Days-2017 which was organized by the Hydrology and Quantitative Water Management Group (HWM) and the Soil Geography and Landscape Group (SGL), at Wageningen University and Research, Publ. of the Netherlands Centre for Riverstudies (NCR) 36-2012, pp. 22-24. http://www.ncr-web.org/ncr-days/ncr-book-of-abstracts.

**Mulatu**, C. A., Crosato, A., & Mynett, A., & Moges, M. M (2105). Effects of dam construction on the planimetric changes of downstream rivers. E-proceedings of the 36th IAHR World Congress, 28 June – 3 July 2015, The Hague, the Netherlands

P. Boeriu , D. Roelvink , **C.A. Mulatu** , C. N. Thilakasiri , A. Moldovanu , M. Margaritescu. (2011): Proceedings of International Conference On Innovations, Recent Trends And Challenges In Mechatronics, Mechanical Engineering And New High-Tech Products Development– MECAHITECH'11, vol. 3, "Modelling the Flushing Process of Reservoirs" page 240 -252.

Netherlands Research School for the
Socio-Economic and Natural Sciences of the Environment

# D I P L O M A

## for specialised PhD training

The Netherlands research school for the
Socio-Economic and Natural Sciences of the Environment
(SENSE) declares that

# Chalachew Abebe Mulatu

born on 29 September 1977 in Dangila, Ethiopia

has successfully fulfilled all requirements of the
educational PhD programme of SENSE.

Delft, 16 December 2021

Chair of the SENSE board

Prof. dr. Martin Wassen

The SENSE Director

Prof. Philipp Pattberg

The SENSE Research School has been accredited by the Royal Netherlands Academy of Arts and Sciences (KNAW)

K O N I N K L I J K E   N E D E R L A N D S E
A K A D E M I E   V A N   W E T E N S C H A P P E N

The SENSE Research School declares that Chalachew Abebe Mulatu has successfully fulfilled all requirements of the educational PhD programme of SENSE with a work load of 30.9 EC, including the following activities:

**SENSE PhD Courses**

- Environmental research in context (2015)
- Research in context activity: 'Co-organizing fifth International Nile Symposium on "Challenges of Blue Nile River Basin and Mitigation Approaches", 5-6 May 2018, Bahir Dar University, Ethiopia'

**Other PhD and Advanced MSc Courses**

- Open Source GIS , IHE Delft (2015)
- Morphological modeling using Delft3D, IHE Delft (2017)
- How to write Scientific Paper (and get it published), TU Delft (2017)
- River Morphodynamics at UNESCO-IHE: Master class, IHE Delft (2017)
- Managing the Academic Publication Review Process, TU Delft (2018)

**Management and Didactic Skills Training**

- Teaching in the BSc course 'Dam Safety and Instrumentation' (2016)
- Teaching in the MSc course 'Hydraulic Structures-II' (2017)
- Supervising MSc student with thesis entitled 'Estimation of Ribb Dam Catchment Sediment Yield and Reservoir Effective Life Using SWAT Model & Empirical Methods' (2020)

**Oral Presentations**

- *Morphodynamic trends of the Ribb River, Ethiopia, prior to dam construction*. IHE Delft PhD Symposium, 2 October 2018, Delft, The Netherlands
- *Flood attenuation and its ecological impact due to upstream dam construction, the case of Ribb Dam, Upper Blue Nile Basin, Ethiopia*. IHE Delft PhD Symposium, 7 October 2020, Delft, Netherlands

SENSE coordinator PhD education

Dr. ir. Peter Vermeulen

9781032250311